ELECTROCORROSION AND PROTECTION OF METALS

ELECTROCORROSION AND PROTECTION OF METALS

GENERAL APPROACH WITH PARTICULAR CONSIDERATION TO ELECTROCHEMICAL PLANTS

JOSEPH RISKIN

ELSEVIER

Amsterdam • Boston • Heidelberg • London • New York • Oxford
Paris • San Diego • San Francisco • Sydney • Tokyo

Elsevier
The Boulevard, Langford Lane, Kidlington, Oxford OX5 1GB, UK
Radarweg 29, PO Box 211, 1000 AE Amsterdam, The Netherlands

First edition 2008

Library of Congress Cataloging-in-Publication Data
Riskin, Joseph.
    Electrocorrosion and protection of metals : general approach with particular consideration to electrochemical
plants / Joseph Riskin. — 1st ed.
      p. cm.
    Includes bibliographical references.
    ISBN 978-0-444-53295-4
    1.   Corrosion and anti-corrosives.   2.   Electrolytic corrosion.   I.   Title
    TA462.R488 2008
    620.1'623—dc22   200841789

British Library Cataloguing-in-Publication Data
A catalogue record for this book is available from the British Library

ISBN: 978-0-444-53295-4

For information on all Elsevier publications
visit our website at books.elsevier.com

Printed in the Great Britain
08 09 10 11   10 9 8 7 6 5 4 3 2 1

In memory of my father
Martin Fruman, *who perished*
in a Soviet concentration camp

CONTENTS

PREFACE

The present book may be considered as a first attempt to regard as a whole the problems of corrosion and protection of metals under the attack of external currents in water-containing environments. Corrosion processes in melted environments are not included in this book, since they are characterized by quite other mechanisms and laws.

Corrosion of metals under the attack of external currents is usually defined by the term "electrocorrosion." This general term of generalizing character is given in the title of the book. However, in the text it is not used too often, as specific questions connected with corrosion under the attack of stray currents or of leakage currents are considered. Besides, in the text it was necessary to specify additional current characteristics (anodic, cathodic, variable, etc).

Based on the general approach to the problems of electrocorrosion and protection of metals, the main focus attention was concentrated on the results of the researches of metal electrocorrosion in aggressive media of electrochemical plants, where these problems were most acute and least studied. These results formed the basis for developing methods and means of protection of metallic equipment and pipelines from corrosive attack by external currents in these plants. The well-known problems of corrosion and protection of metals in the field of stray currents are given in the form of a review.

Alongside the available literary data are given the results of long-term work carried out by a specialized division on the protection of metals against electrocorrosion, headed by the author, in the Moscow R&D Corrosion Institute. These results provide the main content, since available data on the electrocorrosion of metals in aggressive environments are rather limited. It must be noted that all the activities of research and development activities were carried out in constant close contact and cooperation with different operating electrochemical plants of chemical and non-ferrous metallurgical industries.

The successful solution to many problems connected with the design and calculation of models of metallic structures operating in the field of external currents and with researches of corrosion and hydrogenation of metals at cathodic polarization in many respects, is credited to the fruitful cooperation of Dr. J. B. Skuratnik, a brilliant scientist of the Moscow Scientific–Research Institute of physico-chemical researches in the name of Karpov.

The author thanks all colleagues with whom he has had the fortune to collaborate and without whose teamwork this book could not have been born.

The author has pleasure in expressing his special gratitude to the director of the Moscow R&D Corrosion Institute, Professor V. A. Timonin, who, for many years, encouraged this work. His benevolent attitude to the work, professional recommendations and advice given at discussions on the results of the studies promoted a successful solution to many problems.

INTRODUCTION

CORROSION AND ELECTRIC CURRENT – 200 YEARS TOGETHER

The first electric current source was invented 200 years ago, at the boundary of the eighteenth and nineteenth centuries. In 1800, Volta built the famous "column" pile, the first electric battery: a series of alternating copper and zinc disks, separated by cardboard disks soaked with acid or salt solution. Between the copper–zinc pairs, an electric current was generated as a result of zinc dissolution or, in other words, as a result of zinc corrosion. Since then the two phenomena, corrosion and electric current, are closely allied and achievements in the studies of one of them promoted successes in the development of the other.

Nevertheless, only in the third decade of the twentieth century, after publications by Evans and Hoar, did the idea of the electrochemical nature of the corrosion processes of metals in contact with electrolytes finally become established [1].

The development of direct current sources made it possible to use an electric current as a means of influencing the metal–electrolyte system in further investigations. The laws of electrolysis that were discovered by Faraday as a result of these investigations promoted intensive development of theoretical and applied electrochemistry.

The fact that corrosion processes of metals have an electrochemical nature turned corrosion science into a branch of electrochemistry. Methods that are used for studies of corrosion processes are, to a large extent, similar to the electrochemical methods of investigation. Most of these methods (different kinds of polarization measurements, potentiometry, coulometry, impedance, etc.) are based on the influence of external currents on metals. The results of investigations of metal dissolution or deposition kinetics, adsorption and desorption processes, etc., obtained by electrochemists, are of interest to corrosion specialists and vice versa.

Processes based on electrochemical reactions in which electric current plays a part found widespread application in applied electrochemistry (electrolysis processes). The first electrolysis process was used in electrodepositing, the process that is the opposite of the corrosion process. Further investigations led to the development of electrolysis processes without metal deposition in aqueous electrolytes. Among them are water electrolysis for producing hydrogen and oxygen and electrolysis of sodium salts and other chloride solutions for producing chlorine, alkali and oxygen–chlorine compounds. Not only chlorine and oxygen–chlorine compounds, but many other oxidizers: hydrogen peroxide, manganese compounds, persulfates and other peroxides are also produced by electrolysis.

Processes of electrolytic refining and extraction of copper, nickel, cobalt, zinc, chromium, metals of the platinum group and of many other metals were developed.

Some of the metals (aluminum, magnesium, alkali and alkaline–earth metals) are produced by the electrolysis of melted salts.

Electrolysis processes are also applied for organic and inorganic electrosyntheses.

For a number of reasons, including the relative ease of execution and monitoring of electrochemical processes, electrochemical plants are now amongst the leading enterprises in the chemical industry [2].

It was possible to create large-scale and energy-consuming enterprises owing to the appearance of high-power sources of electric current. At some of such plants, current magnitudes attain hundreds of kiloamperes.

The use of electric current in electrochemical plants gave rise to problems connected with corrosion effects of leakage currents penetrating to metallic piping and equipment from electrolytic cells. This became a major problem for many plants, since it hindered the further development of the electrochemical plants. It was very difficult to find corrosion-resistant structural metallic materials for such plants, since these materials were attacked not only by media of high aggressiveness, but also by external currents.

Thus, corrosion phenomena promoted the development of different branches of science and technology connected with the processes stimulated by electric currents. In turn, extensive development of these processes engendered serious problems connected with the corrosion of metals under attack by leakage currents appearing in these processes.

So, we have a two-century long history of studies of the relationship between corrosion and electric current, and there are a huge number of investigations connected with the influence of an electric current on metal–electrolyte systems. It is amazing, but, in spite of these facts, the number of researches devoted to studying the direct corrosion attack of metals by external currents in aggressive environments is rather limited. Even fewer are the number of works devoted to methods and means of metal protection against corrosion by external currents in the aggressive media of electrochemical plants where the action of external currents on the metal structures is unavoidable.

Considered the corrosion problems connected with the attack by stray currents on underground and underwater piping and other structures can be considered an exception. Extensive investigations have been carried out in this field. Methods of control and protection of metals used in the said structures against corrosion attack by stray currents were developed. There is a rich library of books containing chapters on the corrosion of underground structures by stray currents. Therefore, only a limited review on this subject is given in Chapter 2 of this book. Such a review was necessary not only to enlarge the scope of the problems connected with the electrocorrosion of metals, but also to analyze the extent to which the experience that was accumulated in the field of corrosion and protection of metals against the attack by stray currents may be appropriated for solving the problems of electrocorrosion in the aggressive media of electrochemical plants.

The relatively low aggressiveness levels of soil, fresh water and seawater enables the application of carbon steel as a major structural material in these media. Extensive and comprehensive investigations of carbon steels that usually have rather similar corrosion and electrochemical characteristics in these media made it possible to unify most methods and means of protection of underground piping and other structures against corrosion by stray currents. For engineers who are designing and carrying out corrosion protection of these objects today and are using existing standards for these purposes, knowledge in the field of electrical engineering is often more important than knowledge in the field of corrosion and protection of metals.

In electrochemical plants, the situation is quite different since aggressive media and applied structural materials are characterized by their great variety. This variety must be taken into account when the corrosion of metals by external currents and methods of protection against their attack are considered.

This approach to the problem means that a profound analysis of the metal state (passive, active, etc.) in the given aggressive environment of corrosion and electrochemical characteristics (corrosion and activation potentials of the metal, oxidation or reduction potentials of medium components on the metal surface, ratio of external current density to the current density of metal in the passive state, polarization characteristics, etc.) has to be carried out. Under such an approach, the external current is considered to be not just one of the factors of environmental aggressiveness, but the most aggressive one.

A very important point in this approach involves the fact that in the majority of cases, the metals that are used in aggressive media are initially in a passive state. Concentrating on this fact enabled the elaboration of fundamentally new very effective methods and means of protection of metals from electrocorrosion in aggressive media. The major concept of these methods consists of retaining by all means the potential of the protected metal within the limits of its passive field.

Further, it will be shown that such an approach turned out to be the most fruitful, particularly for the development of effective methods of metal protection against electrocorrosion in electrochemical plants.

REFERENCES

1. U. R. Evans and T. P. Hoar, The Velocity of Corrosion from the Electrochemical Standpoint, Proceedings of the Royal Society of London, No. 137, p. 343, 1932.
2. J. O'M Bockris, Electrochemistry for a future society, In: Trends in Electrochemistry, Ed. by J. O'M. Bockris, D. A. J. R. and B. J. Welch, Plenum Press, New York, London, 1977, pp. 1–23.

CHAPTER 1

Dependence of the Corrosion Behavior of Metals Attacked by an External Current on Their Initial State

Contents

1.1. STATE OF METALS IN AGGRESSIVE MEDIA IN THE ABSENCE OF ATTACK BY EXTERNAL CURRENTS

When a pure metal comes into contact with an aqueous solution of its salt, an equilibrium electrode potential, E_{eq}, is established on the metal. This potential corresponds to the equality of speeds of metal atom ionization and of metal ion discharge on its surface. The value of the equilibrium potential of a metal in a 1 N solution of its salt at a temperature of 25°C is identified as the standard electrode potential, which is considered an important metal characteristic. To compare the thermodynamic stabilities of various metals, a so-called electrochemical series was established in which the metals are located in ascending order of their stability. In general, the more positive the standard potential of a metal, the higher is its thermodynamic stability. Values of standard electrode potentials are specified with respect to the standard potential of a hydrogen electrode, which is conventionally accepted as equal to zero.

During electrochemical corrosion, the process of metal dissolution in the electrolytes, which proceeds in accordance with the electrochemical mechanism, two conjugated reactions take place: anodic reaction of metal ionization

$$Me = Me^{n+} + ne^-, \tag{1.1}$$

and cathodic reaction of reduction of any solution component on the metal surface

$$Ox + ne^- = Red. \qquad (1.2)$$

The spontaneous course of a corrosion process according to reaction (1.1) is possible only if the electrode potential of the metal is more negative than the potential of the cathodic reaction (1.2). Such a relationship between the potentials is a thermodynamic condition for the occurrence of the corrosion process. During the course of an irreversible corrosion process, a compromise potential, E_s, is established on the metal. It is called the stationary or corrosion potential.

The hydrogen ion is the component that is most often reduced on the cathode in acid media. The summarizing reaction of cathodic depolarization of a hydrogen ion is characterized by the equation

$$2H^+H_2O + 2e^- = H_2 + 2H_2O. \qquad (1.3)$$

In many media, particularly in water, dissolved oxygen is reduced on the cathode. The summarizing reaction of this reduction is as follows:

$$O_2 + 2H_2O + 4e^- = 4OH^-. \qquad (1.4)$$

The reversible potential of this reaction in a neutral solution in an atmosphere of air (i.e., at a partial oxygen pressure of 0.21 atm) is equal to 0.805 V.

The agent (oxidizer) that is reduced on the cathode is defined in the existing corrosion terminology as the depolarizer of the cathodic process. Accordingly, a corrosion process at which a cathodic reaction (1.4) takes place is termed as corrosion with oxygen depolarization.

Dissolved chlorine, ions of metals of the highest valence, for example, Fe^{3+} and Cu^{2+} and so forth can act as oxidizers in corrosion processes.

The condition of the metal, which can be characterized as stable or unstable in a given aggressive medium, depends on the nature of the metal and on the properties of the aggressive medium. Metals of the first group – lithium, potassium and sodium – are the most unstable, being at the left end of the electrochemical series of metals. The electrode potentials values of these metals (−3.045, −2.925 and −2.714 V, respectively) belong to the most negative. On contact with water, their dissolution is accompanied by strong evolution of hydrogen.

Iron is positioned in the middle of the electrochemical series. The standard potential of the reaction $Fe = Fe^{2+} + 2e^-$ is equal to −0.44 V. The standard potential of the reduction of hydrogen ions in water at pH 7 is equal to −0.414 V, which is a slightly more positive than the standard electrode potential of iron. At such a potential value, dissolution of iron in water with hydrogen evolution is theoretically possible. However, owing to the impedance of the discharge of hydrogen ions on iron, hydrogen evolution in the corrosion process of iron in neutral solutions practically does not occur. Corrosion of iron in water and in neutral solutions that do not contain oxidizers occurs with oxygen depolarization, according to reaction (1.4). In conditions of an easy access of air to the surface, the solubility of oxygen in water is low, of the order of 8 mg/l at ambient temperature. Therefore, the corrosion rate of iron is limited by reaction (1.4), which usually depends on the rate of oxygen delivery to the metal surface.

Metals of the platinum group and gold possess the highest standard electrode potentials values: the potentials of palladium, iridium and platinum are equal, respectively, to 0.99, 1 and 1.2 V; and the potential of gold is equal to 1.5 V. These metals exhibit corrosion stability in most types of aggressive media, including neutral media containing oxygen and other oxidizers.

The estimation of the corrosion stability of metals by the magnitude of the standard potential can be considered only as preliminary since it does not consider the properties of the aggressive media and the ability of the majority of metals to form passive layers on their surface which interfere with the development of the corrosion process.

To execute a more complete analysis of the thermodynamic metal state, potential/pH diagrams for metal/water systems (Pourbaix diagrams) are used [1]. These diagrams take into account the thermodynamic balance between the metal, its ions in solution, insoluble products of reactions between the metal and the solutions at various values of metal potential and the pH of the solutions. Thus, alongside the nature of the metal, the properties of the aggressive medium and its interaction with the metal are considered in Pourbaix diagrams.

Areas of the thermodynamic states of the metal, solution and products of their interaction are separated on Pourbaix diagrams by equilibrium curves. As examples, we shall consider simplified diagrams for the equilibrium conditions at which the activity of metal ions or its soluble corrosion products in a solution is equal to 10^{-6}N. Such an activity value can be considered a border: below this value the metal possesses corrosion stability.

A Pourbaix diagram of iron in aerated water solutions at temperature 25°C is shown in Figure 1.1. The horizontal line **a** corresponds to the equilibrium of the reaction $Fe = Fe^{2+} + 2e^-$. Above this line, in the area limited by lines **a** and **b**, Fe^{2+} ions are stable. The area above lines **b** and **c**, where solid corrosion products of iron are stable, was determined by Pourbaix as the area of passivity.

Lines **a**, **b** and **c** divide the diagram into three main areas. The area of thermo-dynamic stability of iron, which was named by Pourbaix as the immunity area, is disposed below lines **a** and **c**. Between lines **a** and **b**, there is an area of iron corrosion.

The dashed lines 1 and 2 mark the area of thermodynamic stability of water. The top line 1 is pertinent to the equilibrium of the water oxidation reaction, and the bottom line 2, to the equilibrium of the water reduction reaction, with hydrogen evolution. The electrode potentials of iron in neutral water, in the area of its corrosion, are disposed in the area corresponding to the thermodynamic stability of water.

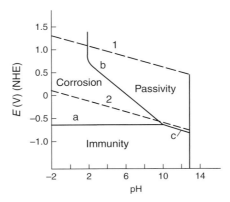

Figure 1.1   Simplified Pourbaix diagram of iron. Temperature, 25°C.

Equilibrium potentials of the iron ionization reaction are, in this area, more negative than the equilibrium potential of the reaction (1.4) of oxygen reduction. This makes the development of a spontaneous process of iron corrosion with oxygen depolarization possible according to reaction (1.4).

The primary products of the reaction of iron with water, ferrous oxide and ferrous hydroxide $Fe(OH)_2$, are oxidized in an aqueous solution to form ferric hydroxide $Fe(OH)_3$, which is known as rust. The layer of this secondary corrosion product practically does not possesses protective properties. At the high concentration of alkali, which has not been indicated in the diagram, soluble products of corrosion – hypoferrites $(HFeO_2^-)$ – are formed.

The simplified diagram given does not consider the influence of widespread aggressive components of the environment, such as chlor-ions. These ions take part in intermediate reactions of the ionization of iron and other metals. In many cases, they stimulate and accelerate the corrosion process. The lines that characterize the ratio of bivalent and trivalent iron ions at various potentials are also not considered in the diagram.

For aluminum, which does not form ions of variable valence and secondary corrosion products, the Pourbaix diagram looks essentially simpler (Figure 1.2).

The area of immunity of aluminum is disposed below lines **a** and **b**. The vertical lines **c** and **d** are pertinent to the equilibrium of the corrosion products of the reactions, respectively:

$$2Al^{3+} + 3H_2O = Al_2O_3 + 6H + 6e^- \tag{1.5}$$

and

$$Al_2O_3 + H_2O = 2AlO_2^- + 2H^+, \tag{1.6}$$

or, in hydrated form:

$$Al^{3+} + 3H_2O = Al(OH)_3 + 3H^+, \tag{1.5'}$$

and

$$Al(OH)_3 = AlO_2^- + H_2O + H^+. \tag{1.6'}$$

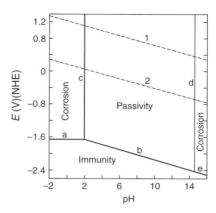

Figure 1.2 Simplified Pourbaix diagram of aluminum. Temperature, 25°C.

Thus, Pourbaix diagrams introduce the concept of the following three states of metals, depending on the potential in the given aggressive medium and the determination of its corrosion behavior: corrosion (active state), passivity (forming passive layers inhibiting the corrosion process on the surface of the metal, including thermodynamically active metals) and immunity (thermodynamic stability) of the metals. Further, it will be shown that these three metal states in aggressive media have a special value when corrosion action by an external current on the metal is considered and when methods and means of protection from electrocorrosion are developed.

1.2. POLARIZATION OF METALS

On the occurrence of an irreversible corrosion process, the ionization rate of the metal prevails over the discharge rate of the metal ions on its surface. The electric neutrality of the metal is retained due to depolarizer reduction by the excess electrons that remain in the metal. The value of the corrosion potential, E_s, that has established on the metal depends on the limiting stage of the corrosion process. If, other things being equal, the anodic process of metal ionization is retarded (which leads to an excess of positive charges in the metal) or the cathodic process of oxidizer reduction is facilitated (which leads to a reduction of the negative charges in the metal), a shift of the metal potential to the positive side takes place. A shift in the potential resulting from the retardation of any stage of the electrochemical process is termed polarization of the metal. The magnitude of this shift, characterizing the degree of difficulty of the proceeding of the electrochemical process, is referred to as overvoltage of this process.

The shift of the potential of the metal to the positive or the negative side is defined, respectively, as anodic or cathodic polarization. When an electric contact of two different metals, for example, copper and zinc, immersed into a common electrolyte occurs, their potentials are drawn together. The potential of zinc, which is negative with respect to the potential of copper, shifts to the positive side, i.e., anodic polarization of the zinc occurs. As a result, the corrosion rate of the zinc increases. At the same time, the potential of the copper shifts to the negative side, i.e., the copper is cathodically polarized. It is possible to measure the current passing between these two metals. In relation to the zinc, this will be an anodic current, and in relation to the copper, cathodic. Thus, the galvanic corrosion of zinc can be considered as a corrosion process proceeding under the conditions of attack by an external anodic current.

Systematic investigations of the corrosion and electrochemical behavior of metals were started in the first half of the twentieth century. The major part of the investigations included analyses of the dependences of the metal potential on the impressed current density (the ratio of current to the working surface area of the metallic sample) in various aggressive media. The dependences determined by polarizing curves provided information on limiting stages and other important details of the mechanism of the corrosion processes; to determine the corrosion and electrochemical characteristics of metals; to estimate the efficiency of electrochemical protection, the influence of the components of the alloys and the danger of galvanic corrosion; and other data connected with corrosion processes. Under the conditions of the action of an external current on metal, the polarizing curves provide direct information on the character, mechanism and intensity of this action.

Polarization curves are usually obtained using glass cells containing three electrodes: the working electrode (the investigated metal specimen), the auxiliary electrode (which is usually made of a platinum wire) and the reference electrode. During the polarization process, the working parts of all three electrodes are immersed into an electrolyte in which the given metal is being investigated. When an elevated purity of electrolyte and high accuracy of the experiments are necessary, the volumes of the auxiliary and reference electrodes are separated from the volume of the working solution of the cell by porous partitions, or these two electrodes are placed into separate vessels connected with the working solution through electrolytic taps. Hydrogen electrodes and other, easier to handle electrodes (calomel, Ag/AgCl, etc.) are used as reference electrodes. The choice of reference electrode depends on the composition and properties of the aggressive environment. The scale of potentials relative to the normal hydrogen electrode (NHE) is accepted as the standard. All the potential values given below are with respect to the NHE.

All three electrodes are connected to an adjustable current source. The power electric circuit, in which the magnitude of the imposed current is measured, is created between the working and auxiliary electrodes. The respective change in the electrode potential with variations in the current density is measured in a circuit created between the metal sample (working electrode) and the reference electrode.

1.3. ATTACK OF EXTERNAL ANODIC CURRENT ON ACTIVELY CORRODING METALS

It is customary to say that a metal that is spontaneously dissolving (corroding) at an appreciable rate, that it is in an active state. When the current of the anodic direction is impressed on an actively corroding metal, the stream of electrons is directed from this metal to an external circuit. According to Faraday's laws, the quantity of positively charged metal ions transferring into the solution is equivalent to the strength of the anodic current. The higher the external anodic current strength, the stronger the anodic polarization of the metal, i.e., the shift of its potential to the positive side.

If the small magnitudes of external current at which the dependence of the potential on the current density close to the linear are neglected, the dependence of the potential on the current density of an actively corroding metal can be expressed by the following equation:

$$E = E_s + (a + b \log i) \tag{1.7}$$

where E_s is the metal potential in the absence of attack by an external current (open circuit or corrosion potential), E is the metal potential at the given impressed anodic current density i, and a and b are constants.

This curve, which has a linear character in semi-logarithmic coordinates (Figure 1.3) is named a Taffel curve. The initial deviation of the curve from the linear character is mainly connected with self-dissolution of the metal, which, at a small external current, is comparable with the magnitude of this current.

Owing to retardation of the ionization process of the metal, under the action of an external anodic current, the potential of the active metal shifts from its E_s value to the positive side. This potential shift, designated overvoltage of ionization, is relatively small; for a change in the current density by one order of magnitude, it usually makes up only several tens of millivolts. The dependence that corresponds to equation (1.7) can be observed, for example, on iron in solutions of acids and neutral salts, where the iron, in the absence of attack by an external current, is corroding, respectively, with hydrogen and oxygen depolarization.

Another example of active metal dissolution is the corrosion of aluminum in solutions of acids and alkalis.

Dissolution of an active metal by an external anodic current takes place at a time when its self-dissolution can lead to a departure from Faraday's law. Under the influence of an external anodic current, the current of self-dissolution usually decreases [2]. This phenomenon is referred to as the positive difference–effect. In some cases, for example, on aluminum and its alloys in chloride–containing environments, a negative difference–effect takes place [2]. This consists of an increase in the self-dissolution rate of the metal under the action of an anodic current. At a high external current density, when the dissolution rate of the metal is high, it is possible to ignore the self-dissolution rate of the metal, lowered as a result of the positive difference–effect. In contrast, in the case of a

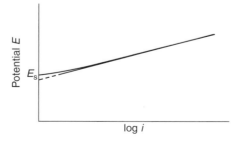

Figure 1.3   Anodic polarization plot of an active metal.

negative difference-effect, the self-dissolution rate can be significant and appreciable deviations from Faraday's law are possible.

The polarization curve of Figure 1.3, in which the current density is an argument and the metal potential is a function, is referred to as galvanostatic, where the metal is maintained at each designated current density up to the establishment of a stationary potential value. It takes a long time to plot galvanostatic curves and to obtain the necessary characteristic potential versus current density values; hence, it is easier to settle for galvanodynamic curves obtained at a higher rate of current density. When plotting galvanodynamic curves, the current density is usually changed at some constant rate and the metal potential at each point of the plot has no time to reach a stationary value. The adjustable current source applied for making such plots is called a galvanostat.

Since the 1960s, corrosion–related electrochemical researches are mainly conducted by means of an electronic electrochemical device named a potentiostat. This device enables the impression of an external current on a metal and its adjustment in such a manner that a designated potential can be supported on the metal. In this case, the potential becomes an argument in the relationship between the potential and the current density. Transformation of the potential into an argument in the study of the dependence of the potential on the current density has a fundamental importance, as the electrochemical state and behavior of metals in an electrolyte are determined, first of all, by their potential values. On the basis of the obtained current density magnitude, the potentiostatic polarization curve characterizes, at each of its points, the metal state at the potential value, which is maintained with the help of the potentiostat.

As potentiodynamic curves are plotted at a certain constant rate, there is no time to establish a stationary value of the current density on the metal. However, in many cases, for revealing the necessary corrosion mechanisms, the results obtained on the basis of potentiodynamic curves, though limited, are enough [3]. Moreover, the commercial potentiostats are able to operate as galvanostats.

It is obvious that the polarization curves obtained on active metals under galvanostatic and potentiostatic operations should coincide, other things being equal.

1.4. ATTACK OF EXTERNAL ANODIC CURRENT ON PASSIVE METALS

Anodic polarization plots obtained on passive metals are characterized by three basic parameters: stationary potential E_s, current density in a passive state, i_p, and potential at the beginning of the current proliferation, E_a (Figure 1.4).

A stationary potential (corrosion potential), E_s, is established on the metal in the absence of external current action. Under anodic potentiostatic polarization of an initially passive metal (i.e., by an artificial shift of the metal potential to the positive side), the magnitude of the anodic current at the beginning is small. As some potential, E_1, is attained, the current density practically stops to vary with the potential variation, or it varies very slowly. This is the current density of the passive state of the metal, i_p (curve 1). However, when potential E_a is reached, the current

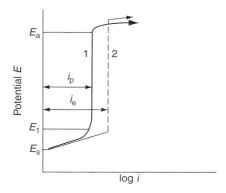

Figure 1.4 Anodic polarization plot of a passive metal: 1 – potentiostatic; 2 – galvanostatic.

density proliferates. This can be attributed to activation of the metal or the beginning of any other anodic process on the metal surface.

Apparently, such a curve cannot be obtained by means of a galvanostat: as the prescribed current i_e exceeds the value i_p ($i_e > i_p$), the potential "falls" to the value which is close to or exceeds the value of the activation potential E_a. Thus, precise values of i_p and E_a cannot be determined by the galvanostatic method. The polarization plots allowing determination of these parameters on a passive metal can be obtained only by using a potentiostatic method, or, at a smaller accuracy, by a potentiodynamic method.

Thus, application of the potentiostatic method of investigation has made it possible to define areas of metal transition from the active to the passive state, as well as other major features of the electrochemical behavior of metals.

One of the most widespread types of corrosion of passive metals is local, or pitting, corrosion. This consists of the formation of damages on the surface of passive metals in the form of pits penetrating into the depth of the metal. In a monograph [4], various theories of the occurrence and development of pitting corrosion are considered. On the basis of the accumulated experimental material, the majority of researchers came to the conclusion that the activation of a passive metal, leading to the development of pitting corrosion, is the result of the attainment by the metal of some critical potential. The value of this potential depends on the presence of activators in the solution. Haloid's ions and, most often, chlor-ions, play the part of activators of pitting corrosion.

As a critical potential, which is also named as the pitting or activation or breakdown potential, E_a, is reached, on some "active centers," where the protective passive layer is most weakened (pores, defects, inclusions, mechanical damages, etc.), a critical concentration of ion–activators is attained. Adsorptive replacement of the passivating agent by ion–activators in the passive layer takes place at this concentration [5, 6]. Being first adsorbed at the active centers, ion activators then form soluble complexes with the metal. Not only does the concentration of the ion activators rise, but the pH value also decreases at the active centers (owing to higher anode currents in these centers and as a result of the hydrolysis of intermediate corrosion products). Therefore, the formed active centers can be maintained for a long time. Under these conditions, the major share of the anodic current (internal current arising between the active centers and the other parts of the metal surface, or external current coming from an external source) will be spent for metal dissolution inside these centers. This causes the formation of local damage – pits on the metal surface. The area between the stationary (corrosion) potential, E_s, and the activation potential, E_a, where the passive state of the metal is maintained (specified by the low and little varying value of the current density, i_p), is referred to as the passive field of the metal.

The other cause of metal activation is its transpassivation – transition of the metal ions to a state of maximal valence. In this case, soluble metal compounds that do not possess protective properties are formed [7]. The phenomenon of transpassivation is observed, in particular, on iron under its anodic polarization in alkaline solutions, which promote the formation of soluble ferrates FeO_4^{2-} containing ions of six-valent iron. The formation of metal ions of maximal valence at the transpassivation potential has been confirmed in investigations on a rotating electrode with a ring of super-austenitic alloys S31254 and S32654 [8].

At the transpassivation potential, local corrosion of metals also develops. The processes of metal activation resulting from pitting and transpassivation (in the absence of ion activators in the solution) are similar in many respects [9]. It is suggested in one work [4] that the breakdown of the passive film in the area of the transpassivation could be connected with the participation of the oxygen of the oxide layer in the process of anodic oxygen evolution.

When the environment does not contain ion activators and the metal is not prone to transpassivation, the potential of some passive metals can be shifted to the positive side by an anodic current, up to values of tens or even hundreds of volts. Metal activation can occur in these cases as a result of the electric breakdown of the passive film.

The factors determining the resistance of passive metals against activation, in particular, against pitting corrosion, in the absence of the action of an external anodic current, and in the presence of such action, essentially differ.

First, let us consider the possibility of the activation of a passive metal in the absence of an external current. If the field of passivity is small, the stationary metal potential can be shifted to the positive side, up to the value of the pitting potential, owing to various causes: contact with another, more positive metal; increase of the concentration of oxygen or of some other oxidizer in the solution; stirring of the solution that contains dissolved oxygen; and so forth.

Increase of the chlor-ion concentration has a slight influence on the value of the metal potential, but narrows the passivity field, owing to a strong shifting of the pitting potential to the negative side. A rise in the temperature also leads to a decrease of the pitting potential. Therefore, metals such as aluminum and stainless steels are successfully applied in solutions with low concentration of chlorides, but are exposed to pitting corrosion at high concentrations of chlorides, especially at elevated temperatures.

On certain areas of the metal, the pitting potential value can be shifted to the negative side, up to the value of the corrosion potential of the metal, as a result of casual mechanical damages on the metal surface, the occurrence of crevice effects (for example, accumulation between the metal surface and a deposited layer of chlor-ions or of other aggressive components destroying the passive film) and so forth. In these cases, when the corrosion potential of the metal becomes equal to its pitting potential, the spontaneous occurrence and development of pitting corrosion occurs on the metal surface. On the polarization curve obtained on the metal which has undergone spontaneous activation, the passive field will be absent and the plot will have a form which is close to the one typical for initially active metals (see Figure 1.3).

The more remote the corrosion potential of the metal from its activation potential, i.e., the longer the passive field of the metal, the less is the probability of its spontaneous activation [10]. Therefore, means of preventing the development of pitting corrosion are focused on increasing the pitting potential by metal alloying, surface passivation or reducing the content of ion-activators in the solution [11]. Activation of the metal can also be prevented by artificial maintenance of the potential of the metal (by means of a potentiostat) at values which are more negative than its activation potential [11].

The magnitude of the current density in the passive state, i_p, is a less meaningful factor of the stability of the passive state of the metal in the absence of attack by an external current. The main role of this parameter is to indicate the stability of the stationary potential of the metal: the higher the value of i_p, the less the fluctuations of E_s. Owing to this, the probability decreases of the potentials E_s and E_a coming closer to each other in the absence of an external current.

As noted, upon attack by an external current of density $i_e > i_p$, the potential of the metal "falls" in the field of active dissolution or the beginning of some other electrochemical process in connection with the anodic oxidation of the solution components on the metal surface (Figure 1.4). Which of these processes – metal activation or solution component oxidation – will proceed, in this case, on the metal surface depends on the properties of the passive film generated on the metal surface and on the relationship between the metal activation potential E_a and the oxidation potential E_{ox} of the solution components. When the passive layer possesses a high electronic conductivity, and the oxidation potential of the solution components on the metal surface is more negative than the activation potential of the metal in the given solution $E_{ox} < E_a$, the metal will not corrode under the attack of an external anodic current. When a non-conductive (barrier to the current of the anodic direction) passive layer is formed, metal activation occurs as a result of the breakdown of this layer. In this case, the stability of the passive layer does not depend on the size of the passive field of the metal – the parameter that determines the metal stability in the absence of the external current.

Thus, in the absence of an external current, the major parameters of metal stability are its activation potential and the size of the passive field. Under the attack of an external anodic current on a passive metal, the major parameters of its stability are the anodic conductivity of the passive layer and the compliance with potential relationship $E_{ox} < E_a$. These parameters are connected with the properties of both the passive metal and the aggressive medium contacting with it.

There might be cases, when the anodic activation potential of the metal is close to, or coincides with, the oxidation potential of the solution components on the metal surface. These cases require special consideration.

1.5. ATTACK OF EXTERNAL ANODIC CURRENT ON THERMODYNAMICALLY STABLE METALS

Platinum and metals of the platinum group possess thermodynamic stability in a wide range of potentials and pH's of solutions. During anodic polarization of these metals, the major anodic process on which the main share of the anodic current is spent consists of the oxidation of the solution components on their surface. In chloride-containing aqueous solutions, the products of the oxidation are, according to the solution composition, oxygen, chlorine and compounds of oxygen and chlorine. It was noted that in a neutral, aerated solution, the equilibrium potential of oxygen evolution, in accordance with the reaction

$$2H_2O = O_2 + 4H^+ + 4e^- \qquad (1.8)$$

is equal to 0.805 V. The value of the equilibrium potential of chlorine evolution by reaction

$$2Cl^- = Cl_2 + 2e^- \qquad (1.9)$$

is more positive; it is equal to 1.33 V. Therefore, under the anodic polarization of platinum, oxygen evolution begins earlier than chlorine evolution. However, platinum is characterized by a high overvoltage of oxygen evolution and by a low overvoltage of chlorine evolution [12]. Owing to this feature of platinum, the quantity of oxygen evolving on its surface is low even at the potential of chlorine evolution, as seen from Figure 1.5. At the same time, at a small potential shift of platinum from the value of the equilibrium reaction (1.9) to the positive side, the rate of chlorine evolution proliferates. At a current density of $1000 \, A/m^2$, only

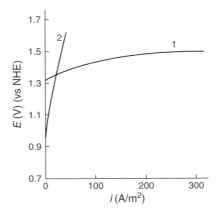

Figure 1.5 Partial anodic polarization plots of evolution on platinum surface: 1 – chlorine; 2 – oxygen [12].

about 1% of the current is spent on oxygen evolution. Thus, the overvoltage value of the anodic oxidation of the solution compounds is an important parameter for thermodynamically stable metals.

Owing to the thermodynamic stability, platinum and metals of the platinum group, particularly iridium, exhibit high corrosion stability under conditions of oxygen and chlorine evolution. Owing to these features, they are still widely used as coatings for commercial anodes in electrochemical plants and in systems of cathodic protection by an impressed current [13].

1.6. ATTACK OF EXTERNAL CATHODIC CURRENT ON METALS

The metal connected to the negative pole of a direct current source receives an excess of electrons which shifts its potential to the negative side. Once the metal attains the reduction potential of any component contained in the electrolyte, a reaction of reduction of this component by the excess electrons becomes possible.

As a rule, a cathodic current decreases the corrosion rate of an actively corroding metal. One of the most widely applied methods of metal protection from corrosion, cathodic protection, is based on this effect [14, 15]. The negative influence of a cathodic current on the cathodic protection of carbon steel is mainly connected with the possibility of scaling of the coat from the surface of a protected metal at a high current density [15]. In weakly acidic media at elevated temperatures, the influence of a cathodic current can initiate corrosion cracking of stressed structures of carbon steel [15].

At potentials of cathodic protection close to $-850\,mV$, austenitic–ferritic stainless steels (duplex and super-duplex) may be prone to hydrogen embrittlement in environments such as sea water [16]. However, when stresses on these steels are not too high, the risk of their embrittlement and cracking is extremely low, as their protective potential in sea water is much more positive (by about $500\,mV$) than the hydrogenation potential [17]. The probability of their embrittlement can be essentially lowered by the prevention of grain growth, which is especially important for welded seams, and by the creation of a structure containing no less than 50% of austenite [16], which is not prone to embrittlement and interferes with the penetration of hydrogen into the ferrite phase.

Valve metals, titanium and tantalum, are prone to embrittlement, which occurs as a result of their hydrogenation under attack by a cathodic current [15, 18].

The influence of an external cathodic current can lead to a reduction of the passive film and to subsequent metal activation. Such an effect, designated as a negative protective effect, is possible, in particular, in the case of cathodic protection of aluminum by an impressed current and also by its contact with magnesium anodes [19]. Corrosion of the aluminum can be attributed in this case not only to the reduction of the oxide film on its surface, but also to the secondary effect of cathodic current influence, which is expressed in alkalization of the solution near the aluminum surface. It is well known that in alkaline solutions, aluminum actively dissolves.

Thus, in general, the influence of a cathodic current on metals that are initially in an active state is to reduce their corrosion rate and is not dangerous for these metals. The influence of a cathodic current on passive metals can also provide protection from corrosion. However, an increase of the protective current above admissible values can lead to corrosion damage of the metals by reduction of the protective oxide layer by solution alkalization near the metal surface or by metal hydrogenation. The sensitivity of passive metals, such as aluminum and titanium, to the influence of the cathodic current is explained, first of all, by their thermo-dynamic instability. These metals are more active than carbon steel. Therefore, their corrosion stability is completely determined by the protective properties of the passive films formed on their surface.

It is important to note, on the basis of the above, that there is a fundamental difference between the cathodic protection mechanisms of initially active and passive metals. Protection of initially active metals, such as carbon steel, in neutral solutions (fresh and sea water, ground, etc.) consists of shifting the metal potential from its stationary value to the value corresponding to its immunity potential (see Figure 1.1). At its stationary potential, unprotected steel dissolves at a significant rate, and for corrosion suppression it is necessary to impress on the metal an external cathodic current which considerably exceeds the magnitude of the corrosion current.

In the cathodic protection of initially passive metals, a totally distinct challenge is posed: to prevent the metal potential from shifting from its stationary value to the value of the activation potential. When the anodic polarization curves of Figure 1.4 were considered, it was noted that when the external anodic current density impressed on the metal exceeds the current density of a passive state ($i_e > i_p$), the potential of the metal shifts to the positive side, to the activation area, and attains a value that is more positive than the value of the activation potential E_a. Conse-quently, when the external cathodic current density impressed on the metal exceeds i_p, the potential of the metal shifts to the negative side and attains a value that is more negative than the stationary potential E_s. The current density of a metal in a passive state, as a rule, is very small. Therefore, the challenge that is posed in the cathodic protection of a passive metal is met by the impression of a cathodic current of a very small magnitude. Moreover, to exclude cathodic activation or hydrogenation of the metal at elevated negative potentials, the magnitude of the cathodic current must be limited.

Researchers of the Norwegian company Corrocean [20, 21] made use of the above-specified feature of cathodic protection of passive metals. They developed a system for the cathodic protection of stainless steel structures in sea water using a cathodic current of a low density. For this purpose they used zinc anodes whose protective current had been sharply lowered by the installation of resistors between the anodes and the protected structures. The current density which was sufficient for the protection of such steels as duplex at room temperature, was in the order of magnitude of 1 mA/m^2. As a result, the service life of the anodes increased, and the number of anodes per unit of length decreased to a great extent. The necessary protective current density for coated carbon steel under the same conditions is equal to about 10 mA/m^2, and for bare carbon steel attains 70 mA/m^2 or more; i.e., for carbon steel, the protection current is higher by about two orders of magnitude than for passive stainless steel [14, 15].

1.7. ATTACK OF EXTERNAL ALTERNATING CURRENT ON METALS

Data on the character of the action of alternating currents on metals are ambiguous. This is explained by the character of the influence of an alternating current which depends not only on its magnitude, but also on such parameters as frequency and amplitude. Moreover, various metals react differently to anodic and cathodic half-cycles of current. For example, it is noted that carbon steel dissolve during the action of anodic, and aluminum during cathodic, half-cycles of the alternating current [22]. Usually, an alternating current renders weaker corrosive attack on metal than direct current. At frequencies of $16\frac{2}{3}$ and 50 Hz, corrosion is observed only at a very high external current density [15]. In ref. [23], it is confirmed that cases of corrosion destruction of underground metal constructions under the action of an alternating current occur extremely seldomly.

Taking into account the initial state of the metals explains much of the resulting data about the influence of an alternating current on metals. As noted, carbon steel is exposed to dissolution under the action of an anodic current, according to Faraday's laws, and a cathodic current renders only a protective effect on it. Hence, this steel dissolves only during the moments of the influence of anodic half-cycles of current and it is protected during the moments of the influence of cathodic half-cycles. Even with such an estimation (without taking into account the after-effect of cathodic protection, which may take place as a result of the cathodic half-cycle action), the corrosion rate of carbon steel under the influence of an alternating current should be less than half that under the influence of a direct anodic current.

Corrosive attack by cathodic half-cycles on aluminum is connected with the above-considered processes of passive film destruction on the metal by cathodic current.

1.8. EXTERNAL CURRENT AS A FACTOR OF THE AGGRESSIVENESS OF THE ENVIRONMENT

The influence of the external anodic current on a metal in an electrolyte solution is usually considered adequate for the creation of an electric contact with another, more electropositive metal, and also for the introduction of an oxidizer into the solution. In both of these cases, shifting of the metal potential to the positive side takes place. When the same potential values are reached under the influence of either of these two kinds of action, a similar corrosion behavior of the metal can be expected as a result. It is obvious that the effect of contact with a more electropositive metal immersed in the same electrolyte is similar to the influence of an external current. It is possible to measure the current passing between these metals and to equate it to the current from an external source. The less obvious analogy of oxidizer influence is proved in ref. [24]. It is shown there in that the dependence of the potential on the current density obtained in 1 N solution of sulfuric acid by potentiostatic polarization and by the introduction of oxidizers of various strengths coincide. In one work [25], additional data on this account are cited, allowing the drawing of generalized conclusions on the role of oxidizers. The specific influence of oxidizers on the formation of passive layer on a metal, for example, upon their inclusion into the structure of this layer, can be considered an exception.

Nevertheless, the specified analogy takes place only within certain limits. Upon contact with other metals, the metal electrode potential is limited by the value of the more electropositive metal. Upon the introduction of an oxidizer into the solution, the potential is determined by the oxidation–reduction potential of the solution containing this oxidizer. Unlike this, there are no limits on the influence of an external current, and the magnitude of the metal potential depends only on the magnitude of the external current. In principle, any high potential can be reached on a metal. Therefore, metals possessing high activation potential value can be subject to corrosion destruction under attack by an external anodic current. Owing to this qualitative difference, the external anodic current should be considered as the most dangerous factor of the aggressive of the environment.

Potential variations at different surface areas of the metal structure, resulting from differential aeration, with distinctions in the surface state and so forth, are less significant. Even though they can cause intensive corrosion damage, their influence is less dangerous than the influence of the external anodic current.

The aggressiveness of the external current is aggravated by its character of non-uniform distribution on the surface of metallic structures and the ability of the current to concentrate on metal areas with a weakened passive film or on defects in the coatings.

The influence of an external cathodic current is similar, to a certain extent, to the effect of contact with a more electronegative metal. Owing to this similarity, protection by an impressed cathodic current and by contact with a more electro-negative material have the common name – cathodic protection.

The reasoning on the limits of the change in the potential of the metal at its contact with another more electropositive metal is appropriate also to the case of metal activation as a result of contact with a more electronegative metal. It is obvious that the potential of the metal at its contact with a more electronegative metal cannot be more negative than the potential of the more electronegative metal. In this case also, the danger of corrosion damage caused by reduction of the passive film on the metal or by its hydrogenation, as a result of attack by an external cathodic current of unlimited magnitude, is considerably higher than which results from contact with a more electronegative metal.

REFERENCES

1. M. Pourbaix Atlas of Electrochemical Equilibria in Aqueous Solutions, Pergamon Press, New York, 1966.
2. N. D. Tomashov, Theory of Corrosion and Protection of Metals, The Macmillan Company, New York, 1966, 672pp.
3. N. D. Tomashov and G. P. Chernova, Passivity and Protection of Metals Against Corrosion, Plenum Press, New York, 1967, 208pp.
4. Z. Szklarska-Smialovska, Pitting and Crevice Corrosion, NACE, Houston, 2005, 650p.

5. Ja. M. Kolotirkin, Khimicheskaya Promishlennost' (in Russian), No. 9, 1963, pp. 38–46.
6. L. I. Freiman, Stability and Kinetics of Pits Development (in Russian), Review "Itogi Nauki I Tekhniki, Korroziya I Zaschita Ot Korrozii", Vol. 11, VINITI, Moscow, 1985, pp. 3–71.
7. H. H. Uhlig and R. W. Revie, Corrosion and Corrosion Control, John Wiley & Sons, New York, 1985, 464pp.
8. I. Betova, M. Bojinov, P. Kinnunen et al., J. Electrochem. Soc., Vol. 149, No. 11, 2002, B499–B509.
9. D. Landolt, Transpassivity, In: Collected Articles: Passivity of Metals, Ed. by R.P. Frankentahl and J. Kruger, The Electrochemical Society, Inc., Princeton, NJ, 1978, pp. 484–504.
10. P.H. Schweitzer Ed., Encyclopedia of Corrosion Technology, Marcel Dekker, Inc, New York, 1998.
11. J. Kruger, Passivity, Hulig's Corrosion Handbook, 2nd edn, by R.W. Revie, Wiley & Sons, New York, 2000, pp. 165–171.
12. N. P. Fedotiev, A. F. Alabishev, A. P. Rotinian et al., Applied Electrochemistry (in Russian), Khimiya, Leningrad, 1967, 600pp.
13. L. M. Yakimenko, S. D. Khodkevitch, E. K. Spasskaya et al., Platinum–Titanium Anodes in Applied Electrochemistry (in Russian), Review "Itogi Nauki I Tekhniki, Electrokhimiya", Vol. 20, VINITI, Moscow, 1982, pp. 112–152.
14. J. Morgan, Cathodic Protection. NACE , 1987, 519pp.
15. W. V. Baeckmann, W. Schwenk and W. Prinz (authors and eds), Handbook of Cathodic Protection: Theory and Practice of Electrochemical Protection Processes, Ed. by W. V. Baeckmann, W. Schwenk and W. Prinz, 3rd edn, Gulf Publishing Co, Houston, 1997, 520pp.
16. R. Francis, G. Byrne and G. R. Warburton, Corrosion, Vol. 53, No. 3, 1997, pp. 234–240.
17. C. P. Dillon, Corrosion Resistance of Stainless Steels, Ed. by Ph. A. Schweitzer, Marcel Dekker, Inc., New York, 1995, 365pp.
18. J. R. Davis, Ed., Corrosion, Understanding the Basics, ASM International, Materials Park, OH, 2000, p. 258.
19. M. Cherni, Zashchita Metallov, Vol. 11, No. 6, 1975, pp. 687–698.
20. R. Johnsen, P. O. Gartland, S. Valen and J. M. Drugli, Materials Performance, Vol. 35, No. 3, 1996, pp. 17–21.
21. J. M. Drugli, Method and Arrangement to Hinder Local Corrosion, and Galvanic Corrosion in Connection with Stainless Steels and Other Passive Materials, Patent WO 92/1667, Norway, March 1991.
22. U. R. Evans, The Corrosion and Oxidation of Metals: Scientific Principles and Practical Applications, Edward Arnold Ltd, London, 1960.
23. R. Marchal and G. Poirier, Corrosivity of Soils to Ferrous Metals, In: Water and Gas Mains Corrosion, Degradation and Protection, Ed. by C. Basalo, Ellis Horwood, New York, 1992, pp. 149–182.
24. N. Ya. Bune and G. M. Kolotirkin, Jurn. Phiz. Khimiyi (in Russian), Vol. 35, No. 7, 1961, pp. 1543–1550.
25. Ya. M. Kolotirkin, G. M. Florianovich, and A. I. Kasperovich, Using of Oxygen for Protection of Structural Metallic Materials from Corrosion in Aqueous Media, Review "Itogi Nauki I Tekhniki, Korroziya I Zaschita Ot Korrozii", Vol. 8, VINITI, Moscow, 1981, pp. 3–5.

CORROSION AND PROTECTION OF UNDERGROUND AND UNDERWATER STRUCTURES ATTACKED BY STRAY CURRENTS

Contents

2.1. MAIN MEDIA, SOURCES OF STRAY CURRENTS AND OBJECTS OF THEIR CORROSIVE ATTACK

Electrocorrosion and the protection of metals under the conditions of attack by the external currents named stray currents on underground pipelines and constructions, have been thoroughly studied. A vast amount of scientific knowledge and practical experience has accumulated in this field [1, 2].

Soil is the most widespread environment for attack on metals by stray currents. There are in the soil thousands of kilometers of pipelines for the most varied purposes: underground parts of gas and oil facilities, metal and reinforced concrete foundations of buildings and constructions, supports of power transmission lines, etc.

The most typical pH values of soil are close to neutral, but can vary from acid (pH 4.5–5.5) to alkaline (pH 7.5–8.5 and more). Acidic soils, below pH 4.5, and alkaline soils, above pH 8.5, are seldom met, and are a result, most probably, of pollution by waste generated by some manufacturers.

Carbon steel is the basic structural metallic material for underground pipelines and constructions. As noted in the previous chapter, in environments that are close to neutral, with access to air, carbon steel is in an active state and corrodes with oxygen depolarization.

For underground constructions, general non-uniform corrosion is typical. In the absence of an external current, its average rate is nearly 0.2–0.4 mm/year [3]. Factors such as the formation of macro-couples of differential aerations and concentrations (owing to unequal access to air at different parts of the constructions, or to differences in the composition and structure of the soils), crevice corrosion and the activity of various types of bacteria, strengthen the corrosion aggressiveness of soils. In addition, owing to the influence of the specified factors, corrosion can get a localised character.

In this connection, regardless of the possible influence of stray currents, underground pipelines and constructions are supplied with means for protection against soil corrosion. The most widespread means of protection are anticorrosive coatings and cathodic protection. These can be applied separately or in combination (combined protection).

The sources of stray currents are the lines of underground trams and electrified railways, as well as electric welding devices, installations of cathodic protection and other electric installations. Spreading in the ground, currents create electric fields at distances of tens of kilometers from their source. Therefore, they have received the name stray currents, reflecting the arbitrary character of their distribution over great distances.

Factors of aggressiveness, such as acidity, the presence of oxidizers or activators and temperature, are considered when choosing materials and ways of protection, as these characteristics are usually known in advance. In contrast, attack by stray currents on the metal structure is not always expected, and consequently, it is not always taken into account.

An illustration of stray current penetration to an underground pipeline from a railway or tram power line is given in Figure 2.1. In these lines, the rails are part of an electric circuit. At rail joints, where their electrical resistance is great, a part of the current flows down into the ground, choosing those directions with the least electrical resistance. Metallic pipelines and constructions in the ground are the best routes for current flow. When underground pipelines and constructions are insulated from the ground by means of various kinds of coatings, stray currents penetrate to the metallic pipelines and constructions through pores and defects in these coatings.

Currents have a cathodic direction in the zones of penetration to the pipeline. In other zones, where there are also defects in the coating or the resistance between

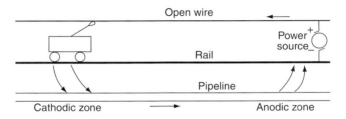

Figure 2.1   Scheme of attack of an underground pipeline by stray currents.

the metal and the ground is lowered for any reason, currents flow down from the metal into the ground. In these zones, currents have an anodic direction.

As stray currents are one of the most dangerous factors causing corrosion damage to metals, methods for their detection and control, as well as means of protection of constructions from their attack, are usually standardized and regulated [4].

The factors defining the corrosion rate of metals in seawater, in the absence of the influence of stray currents, are divided into chemical, physical and biological [5]. The content of dissolved oxygen and carbon dioxide, salinity (first of all, the content of chlorides) and pH of seawater comprise the chemical factors. The basic physical factors are electric conductivity, flow rate and temperature of the water. The biological factors influencing corrosion processes are concerned with the biological activity of living organisms, which is connected with absorption or evolution by these organisms of oxygen and carbonic acid, which can lead to metal fouling. All the specified factors are closely interconnected. pH values in the surface layers of seawater change within narrower limits than in the ground and are usually within the range of 7.5–8.5. The average value of water salinity in the three oceans (Pacific, Atlantic and Indian) and also in the seas adjacent to them is close to 35‰ (35 g in 1000 g of water). At such a salinity value, the content of chlor-ions in seawater reaches 20 g/kg. In more distant seas, the salinity of the water differs essentially from this value. Thus, in the Baltic Sea it is equal to 8‰, and in the Caspian Sea, 13‰. Owing to the high content of mineral salts in seawater, its conductivity is high, reaching a value of $0.05\,\Omega^{-1}\text{cm}^{-1}$, in ocean water at 20°C.

In seawater, especially under intensive stirring, the conditions for the removal of corrosion products from the metal surface are more favorable than in ground. Owing to this, as well as the high conductivity and high content of chlorides, the corrosion rate of carbon steel in seawater is generally (in the absence of stray currents) greater than the corrosion rate in the ground.

The occurrence of stray currents in seawater is much less probable than in the ground. A vessel can undergo corrosion by stray currents during welding work when the welder, a source of direct current, is located on the coast, its positive pole is connected to the vessel and its negative pole is connected to the welding rod [6, 7]. The influence of stray currents is especially dangerous when two or more vessels located near the coast are connected to such a circuit. As a stray current of significant magnitude passes between the vessels, this leads to the occurrence difference in the potential between them. If there are defects in the coating of the vessel's hull, from where the current drains off into the water, a current of high density concentrates in these defects and it causes an intensive corrosion of a local character. Owing to the large cross-section area of the water interposed between these two vessels and the high conductivity of seawater, it is impossible to prevent the occurrence of stray currents even by connecting the vessels with a copper wire with a large cross-section.

Owing to the relatively low conductivity of fresh water, stray currents that can occur in fresh water from the same causes as in the sea are less dangerous, other things being equal.

The most extensively used concrete, prepared on the basis of Portland cement with carbon steel reinforcement, has a high alkalinity (pH 13). At such pH values, the carbon steel reinforcement has a passive state in this concrete [8].

In concrete, the humidity of which corresponds to the humidity of air, the steel reinforcement is minimally subject to corrosion in the absence of attack by an external current attack.

Structures of reinforced concrete are applied in a wide range of objects that can be subjected to attack by stray currents. Amongst them are elements of buildings and constructions of electrochemical plants, supports of power transmission lines and other structures in contact with sources and conductors of electric current.

2.2.   METHODS OF DETECTION AND CONTROL OF STRAY CURRENTS

Methods of current detection on pipelines and other underground metallic structures, to a significant extent, are based on measurements of voltage difference between the structure and the ground [1]. The potential of the metal is measured with respect to a copper–copper sulfate reference electrode. The latest one consists of a cylindrical vessel filled with a saturated solution of copper sulfate into which a core of pure copper is immersed. At the bottom part of the vessel, there is a porous partition providing electric contact of the solution with the ground by pressing the electrode against the ground surface during the measurements. Millivoltmeters with a high input resistance are applied for the measurements.

The average values of the potentials of metals measured against the copper–copper sulfate reference electrode in the ground, in the absence of attack by an external current, are as following: for carbon steel −0.55 V, for lead −0.48 V and for aluminum −0.7 V.

On the basis of numerous measurements, the average values of the metal potentials of a structure are determined. Then, the difference can be found between the said average value and the metal potential in the ground at each measurement point. This allows the plotting of diagrams of the potential variation along the structure. Steady values of potential difference indicate the absence of stray currents. If the measured values of the potential difference varying by magnitude, and, especially, by magnitude and by polarity, this testifies to the presence of stray currents in the ground. The most dangerous zones of stray current influence are disposed in the field of jumps in the potential values. In these zones, the measured potential values are significantly more positive than their average values found in the absence of stray currents. Thus, the presence of anodic and sign-variable zones on underground metallic pipelines and other structures is taken to be a criterion of corrosion danger caused by stray currents.

In the absence of extended metallic constructions in the ground, the detection of stray currents is carried out by measurements of the voltage difference between two points in the ground which are spaced 100 m apart. These measurements are made with the help of copper–copper sulfate reference electrodes placed every 1000 m in mutually perpendicular directions. For these measurements, millivoltmeters of high input resistance are also used.

Measurements of the potential shift of the metal along a construction in the areas of alternating current attack are also carried out with the help of copper–copper sulfate reference electrodes and magneto–electric system devices. In doing so, the stationary potential of the metal with respect to the reference electrode is balanced out by the inclusion of a counter electromotive force into the circuit.

Measurements of the current strength and of the current direction on the extended pipeline and other constructions are carried out by the use of high-sensitivity millivoltmeters with scales of 1 and 10 mV. The distances between the contact points of the millivoltmeter with the construction are 200–400 m, depending on the distance between the control points or inspection chambers that are located along the construction. Using the measured data on the voltage drop at a given section, the resistance of which is known or can be calculated, the strength of the current flowing along the construction can be calculated using Ohm's law.

Owing to the labor input needed for works on the detection and control of stray current values, monitoring systems making use of zinc and other types of reference electrodes incorporated into the ground along the construction were proposed in Ref. [9]. Application of such systems demands significant expenses, but essentially reduces further labor input and raises the control reliability.

The magnitude of the external anodic current density attacking the metal at some spot of current, draining off to the ground through a defect in the metal coating, depends on the following factors: the magnitude of the current that penetrates the construction, the character of the current distribution along the construction, the general number of defects in the coating and the surface area of the given defect. There are also other factors that cannot be controlled.

In the city, branching in the city conditions of the rail network, which consists of interconnected closed and open contours, with much transport moving amongst them at varying speeds and loadings makes the system rails – ground–underground construction system extremely complex. Solving problems of the distribution of potentials and currents in such a system requires using a complex mathematical technique [10].

The danger assessment of corrosion attack by stray currents of bars in reinforced concrete structures is even more problematic [8]. In addition to the above-mentioned factors which complicate this problem, a number of new ones are added. Among them are the following: bars ramification of bars inside the concrete, impossibility of breaking off this reinforcement (which form closed circuits of metallic conductors inside the concrete) to measure stray current values, dependence of current distribution on the humidity of the concrete at different sites of the structure, etc. Therefore, the assessment of corrosion danger of the reinforcement by stray current attack has to be reduced to the revealing of the presence or absence of stray currents in the structure.

In ref. [8] the magnitudes of the stationary (open circuit) potentials of reinforcement–ground and reinforcement–concrete, with respect to a copper–copper sulfate reference electrode, are given. These data were obtained as a result of a wide range of measurements (Table 2.1).

If the measured values of the difference of the potentials fall outside the range specified in Table 2.1, this indicates the presence of stray currents.

Thus, a quantitative estimation of the stray current value and, especially, of the current density that attacks the construction is far from being consistently achievable

Table 2.1 Ranges of difference in potentials of reinforcement–concrete and reinforcement–ground in the absence of stray currents

Construction location	Measuring circuit	Range of difference in potentials, V ($Cu/CuSO_4$)
Above ground	Reinforcement–ground	−0.5 to 0
	Reinforcement–concrete	−0.3 to 0
Underground	Reinforcement–ground	−0.8 to −0.15

in practice. As stray currents are revealed in the ground, it is necessary to undertake all possible measures for the maximal decrease of their magnitude and for the protection of underground constructions made of metal and of reinforced concrete against their corrosion attack.

2.3. METAL CORROSION BY STRAY CURRENTS

Construction areas from which stray currents drain off into the ground (i.e., metal areas attacked by anodic currents) are exposed to corrosion damage. Ions of metal bearing a positive charge are passed into electrolytes, which in the cases considered are ground, water or concrete.

As iron and carbon steel are in ground and in water in an active state, the obtained anodic polarization curves have a form corresponding to the Taffel dependence, illustrated in Figure 1.3. The slope of the curve (i.e., the polarizability of the metal), obtained in ground, with other things being equal, will be steeper than that obtained in seawater. This is explained by the higher seawater conductivity and the accumulation of corrosion products near the metal surface in ground, due to their complicated removal from the metal surface in this environment.

On attack by stray currents of anodic direction, the corrosion rate of carbon steel both in ground and in seawater is defined by the magnitude of the anodic currents at the areas of their attack. Irrespective of the slope of the polarization curve, at potentials that are more negative than the potential of oxygen evolution, practically all the current which is draining off from the metal into the ground or into the water is spent on metal dissolution.

According to Faraday's law, a current flow of 1 A through a steel construction leads to the dissolution of 9.1 kg/year of metal. Values of stray currents reach hundreds of amperes and are usually concentrated at areas of limited size. In this connection, the corrosion by stray currents has a local character and can proceed at very high rates which aggravate the danger of their attack.

In some cases, it is difficult to determine from the character of the local metal corrosion, whether it is the result of the occurrence of macro-couples of differential aeration or the result of a stray-current attack. Extremely high corrosion rates can specify the attack by a stray current. This may exceed, by even ten times, the corrosion rate of carbon steel in ground in the absence of stray currents [10].

In these cases, the soil corrosion can be neglected and only corrosion by stray currents needs to be considered.

The corrosion rate of underground constructions in the areas of alternating current (a.c.) attack is usually lower than that in the areas of direct current (d.c.) attack. The sources of the a.c. are the rail transport working on a.c. and city electric systems with a ground neutral. During a.c. attack on an underground construction of carbon steel, dense corrosion products are formed, which reduces the corrosion rate at the attacked areas. However, an increase of the current density on other metal areas, where the layer of corrosion products was not formed, may lead to the development of local corrosion development along the structure.

The corrosion rate of low-alloyed steels in the absence of stray current attack can be a little lower in some soils than the corrosion rate of carbon steel [3], but the attack by stray current levels off the advantages of these steels owing to the causes mentioned above.

High-alloyed chrome and chrome-nickel stainless steels possess a high stability in soils. However, the presence of chlor-ions, especially in combination with the effect of differential aeration and other factors of soil aggressiveness, can become the cause of local corrosion development on stainless steels in soil. Considering the high cost of the stainless steels, their application as structural materials for underground constructions is avoided.

Stainless steels 18-10 are prone to pitting corrosion in seawater. The penetration rate of pits into a steel that does not contain molybdenum, exceeds 3 mm/year, and with a molybdenum content of 3% decreases to about half that value [11]. In this connection, steels of type 18-8 are inapplicable in seawater. High-alloyed chrome-nickel stainless steels, with the addition of 3% of molybdenum and 0.25% of nitrogen (super-duplex), possess a quite high stability in seawater in the absence of external current attack. However, as noted in Chapter 1, anodic current attack leads to pitting corrosion development, and hydrogen embrittlement of these steels is probable under cathodic protection – especially in zones of welded seams and thermal influence [12].

Aluminum and lead are used as materials for sheaths and armors of underground cables [13]. In the absence of stray currents in ground, the corrosion rate of these metals is significantly less than the corrosion rate of carbon steel, but this strongly depends on the composition and structure of the soil. For instance, with good air access to the metal surface, aluminum and its alloys possess high stability in neutral soils. With poor air access and a high chloride content in the soil, the local corrosion rate of aluminum can exceed the corrosion rate of carbon steel.

Lead sheaths of cables possess the highest stability in environments containing chlorides – especially in the presence of sulfates, carbonates and silicates, which dense salt films on lead. However, in marshy soils with a high content of carbonic acid, the soluble bicarbonate of lead is formed on the lead surface, and the metal loses its corrosion stability.

Under the attack of stray currents on aluminum and lead sheaths of cables, the influence of the structure and properties of the environment become minor, and the corrosion rate, as in the case of steel constructions, is determined by the magnitudes of the currents draining off from the metal into the ground. The influence of anodic currents leads to the dissolution of lead at the boundaries of the grains. Therefore, the occurrence of intercrystalline corrosion of lead sheaths of cables can indicate the presence of stray currents in the ground [6].

As noted, currents of a cathodic direction, which take place at the areas of stray current transfer from the ground to the underground construction, do not exert corrosion attack on carbon steel, but can cause corrosion damage to aluminum sheaths of cables. As is apparent from the potential–pH diagram for aluminum (see Figure 1.2), aluminum oxide, providing a passive state to the metal, maintains stability at the range of values of pH 4–8. Under the influence of cathodic current, as a result of environmental alkalization near the metal surface, the pH value can be considerably above 8 in this zone. The area of aluminum immunity is disposed at potentials that are much more negative than the equilibrium potential of hydrogen evolution. Therefore, intensive corrosion of aluminum, with hydrogen depolarization, is possible when its protective passive film is reduced by a cathodic current.

The influence of a cathodic current on lead is less dangerous than on aluminum, though in this case corrosion of the lead, accompanied with plumbite formation, is possible. This compound hydrolyzes and forms red crystals of lead monoxide. The presence of such crystals on some areas of cables with a lead sheath is an indicator of the influence on these sites of stray currents of a cathodic direction [6].

The a.c. is much less dangerous than the d.c., not only for underground constructions from carbon steel, but also for lead cables. Corrosion losses caused by a.c. at the areas of steel constructions and of sheaths of lead cables are by two orders of magnitude lower than the losses caused by d.c. on these objects, other things being equal. At the same time, attack by a.c. on aluminum sheaths of cables can lead to their intensive corrosion damage. At a current density above $50 \, A/m^2$ the corrosion damage caused by a.c. reaches 50% of the damage caused by d.c. of a similar magnitude [14].

Chloride-containing components are often added to concrete to increase its strength [6, 8]. Under frequent concrete humidification and a shift in its pH to neutral values, corrosion development of the reinforcement in chloride-containing concrete is possible, even in the absence of stray currents attack. In such an environment the carbon steel reinforcement loses its passive state. High humidity and the presence of chlorides raise the conductivity of the concrete and make it especially available to the penetration of stray currents [11]. During their attack on the actively corroding reinforcement, the current is almost completely spent on its dissolution. As a result of intensive corrosion of the reinforcement, fast destruction occurs of the reinforced concrete structures occurs [15].

The volume of the corrosion products more than twice exceeds the volume of the steel which has been transformed into these products by more than twice. Therefore, high internal stresses occur as a result of the corrosion of the reinforcement and this leads to cracking of the concrete and a sharp decrease in the strength of the structure. In addition, an electrolysis process caused inside the concrete by the attack of stray current leads to destruction of the concrete structure. According to data given in ref. [8], destruction of a concrete structure by a current of high density occurs even in dry concrete. At anodic current densities of $3–4 \, mA/cm^2$ on the reinforcement, cracks in concrete appear after 4–5 days and at current densities of $0.2–0.3 \, mA/cm^2$ they appear after 25–30 days.

According to ref. [16], the influence of an external current raises the degree of localization of corrosion damage. For example, after six years of reinforcement testing in humidified concrete containing 3% $CaCl_2$, the ratio of maximal corrosion rate of reinforcement (1.2 mm) to its average corrosion rate (0.26 mm) was equal to ~5. After testing under attack by an external anodic current, this ratio increased to 20.

2.4. PROTECTION OF METALS AGAINST CORROSION ATTACK BY STRAY CURRENTS

2.4.1. Measures for reducing stray currents

Measures for diminishing stray current penetration to underground constructions should be provided at the initial stage of design. Constructions should be located as far as possible from stray currents sources – lines of electrified railway transport and high-voltage power lines. The routes of these constructions should pass through areas of minimal ground humidity and should cross the said lines as little as possible; the angles corners of their crossings should be close to 90°.

According to existing standards, rails should be bedded in broken stone, gravel or other equivalent (regarding their insulating properties) ballast. Wooden ties must be impregnated with non-conductive oil antiseptics. At the time of application of ties of reinforced concrete, it is necessary to insulate them from the rails. The higher the contact resistance between the rails and the ground and the less the longitudinal resistance of the rail joints, the lower are the leakage currents from the railway into the ground. Reduction of the longitudinal resistance is attained by the connection of adjacent joints using flexible copper wire or other conductors.

The most important component of the protection of underground constructions is their insulation from the ground. The higher the construction–ground resistance (i.e., the better the quality of the insulating coating of the construction), the smaller the stray current magnitudes penetrating to the construction. Quality insulation of underground constructions is capable of diminishing by hundreds of times the magnitude of currents reaching the construction.

It is impossible to completely exclude defects in the coatings of extended constructions in practice. Therefore, alongside the application of insulating coatings, the longitudinal resistance of pipelines is increased, to decrease the magnitude of stray currents getting to them [10]. For this purpose, electric sectionalization of pipelines is executed [17]. This consists of the application of insulating flanges and the insertion of insulating material into the pipelines. In this way, it is possible to significantly lower the magnitude of the current on the pipeline. However, along with decrease of the current size, the number of anodic zones, where the current is draining off from the pipeline into the ground, increases. This number is equal to the number of sections breaking the current flow along the pipeline. Therefore, sectionalization is usually carried out together with the application of grounded current taps [18]. Magnesium or zinc anodes can be used as current taps. Along with current tapping, these anodes provide cathodic protection of the pipeline at the areas of current drain off into the ground. In addition, the application of current taps protects pipelines from breakdown of the insulation in the case of high-voltage action – in particular, when lightning strikes. Shunts with variable resistance, as well as diodes providing unidirectional current flow, can be installed between insulating flanges to regulate the magnitude of the current flowing along the pipeline [19].

For the protection of telecommunication and underwater cables bedded on the sea-bottom, grounding anodes mounted at the ends of cables are also applied [20]. The sizes of the current flowing along telecommunication cables are nearly 1 A, and the grounding does not require special safety measures. At the same time, upon the installation of the grounding near power cables, where the current strength attains several hundreds of amperes, it is necessary to locate the adjacent underwater constructions as far as possible from the cable.

Prevention of the occurrence of stray currents that attack the hulls of ships during welding operations is achieved by separation of the electric circuit schemes in which ships and seawater are participating [6].

Special standards exist for determining the stray current reduction measures in the reinforced concrete structures of electrochemical plants of the chemical and metallurgic branches of industry. Electrolyzers and bus ducts are the major stray current sources in these plants. Overlaps, platforms for the maintenance of electrolyzers, columns and beams for supporting bus ducts, as well as underground structures of reinforced concrete, are the objects of the attack by stray currents.

To reduce stray-current penetration to structures of reinforced concrete, standards forbid the application in electrochemical plants of materials that are capable of absorbing moisture (concrete, non-glazed porcelain, ceramics, etc.) without special treatment by water-repellent and insulating compositions.

The reduction of stray currents can be achieved by sectionalization, which in this case consists of insulating seams forming at overlaps, platforms for the maintenance of electrolyzers and underground structures of reinforced concrete. Overlaps for the installation of electrolyzers should be separated by insulation seams from adjoining walls, columns and other elements of the building.

Polymer–concrete compounds have to be applied for constructions that are located close to the current sources in electrochemical plants.

To prevent current draining off from the reinforcement into the concrete, it is necessary to insulate concrete foundations by painting, by insulating coatings or by using insulating concretes.

For specific purposes, it can be of interest to apply polymer–lime-concrete compound with the addition of 0.5–0.7% furfuryl alcohol. This material possesses the ability of self-packing on contact with an acidic environment [21]. Researchers have shown that when an anodic current is imposed on steel bars immersed into this compound, acidification and self-packing of the concrete layer adjacent to the bar surface take place. As a result, the magnitude of the current that drains off from the into the concrete sharply decreases.

2.4.2. Protection of underground structures by electrodrainage

Electrodrainage protection is the basic method for the protection of underground construction from the corrosion attack by stray currents [1, 7, 10, 17]. It is intended for the elimination of anodic zones from underground constructions by stray-current drainage to its source. On the lines of electrified transport, it is carried out by an electric connection of the underground construction with the negative bus of a traction substation or with a section of the rail circuit.

Connection of the underground construction directly to a current source or to the rails is referred to as a direct electric drainage. This is applied when there is no danger of current running off from the rails to the underground construction.

If the potential difference between the construction and the rail is sign-variable or positive, a polarized relay or a valve device is installed on the line of construction connecting to the rail circuit, and current passes through it only in one direction: from the construction to the rails. Such a type of the electric drainage is referred to as polarized drainage.

When there are several current sources in the area of the location of the underground construction, polarized drainage does not provide complete elimination of the anodic zones from this construction. In this case, protection by means of reinforced drainage is carried out, comprising a combination of polarized drainage and an cathodic protection by an impressed current. The negative pole of a direct-current source (a station of cathodic protection) is connected to the underground construction, and the positive pole is connected to the electrified paths, which are the sources of the stray current. Such protection plays the role of polarized drainage since the current is passing in only one direction. When the potential of the railway section reaches a value which is positive with respect to the potential value of the closest part of the protected underground construction, the installation provides cathodic protection to the construction in which the rails serve as the grounding. The shortcomings of reinforced drainage which limit its application are related to the increased wear of the rails used as the grounding and the increased energy consumption for this kind of protection.

2.4.3. Cathodic protection

As carbon steel in ground and in water has an active state, and coatings cannot provide complete protection from corrosion in these environments, corrosion prevention of underground and underwater constructions is usually carried out by cathodic protection, irrespective of the presence or absence of stray currents in the area where the construction is located. At the same time, at the stage of design of the cathodic protection, it is necessary to consider data about the presence of stray currents [7, 22].

Cathodic protection of aluminum sheaths of cables in areas of alternating current attack is inapplicable [23]. Added to the cathodic component of an alternating current, the cathodic protection current can lead to activation and accelerated corrosion of aluminum. The basic method for the protection of aluminum sheaths of cables is their reliable insulation from contact with the environment.

The materials applied for manufacturing grounding anodes in cathodic protection by an impressed current are characterized by their wide variety. Steel scrap is sometimes used for this purpose, since its application is attractive because of its low cost [7]. Steel anodes dissolution at current outputs is close to 100%. In this connection, steel grounding anodes, designed for a long-term service life, should have a maximal weight.

Graphite is a fairly inexpensive and stable anode material. Its major limitation is its low mechanical strength.

Grounding anodes made from silicon cast iron, fairly inexpensive material comprising an alloy of iron with $\sim 15\%$ silicon, are the most widely applied [7]. Because of the formation of a silicon dioxide passive film on its surface, this alloy maintains a very high stability under the conditions of anodic polarization at current densities up to $50\,A/m^2$. The anodic current is mainly spent on this material for oxygen evolution. However, during chlorine evolution the passive film of the silicon cast iron fails, and it starts to dissolve actively. Therefore, anodes of silicon cast iron are applied in environments with a low content of chlorides, such as soil and fresh water.

When grounding anode's for high current densities are required, titanium anodes with coatings based on platinum group metals (platinum, iridium) [7, 24] and on metal-oxide compounds, such as ruthenium/titanium oxides [25], and ceramic mixed metal oxide (MMO) anodes are applied [26].

The properties of anode materials are considered in more detail in Chapter 11.

REFERENCES

1. B. G. Volkov, N. I. Tiossov and V. V. Shuvalov, Handbook on Protection of Underground Metallic Constructions from Corrosion (in Russian), Nedra, Leningrad, 1972, 224p.
2. M. Romanoff, Underground Corrosion, 2nd ed., Houston, NACE, 1989, 227p.
3. N. D. Tomashov, Theory of Corrosion and Protection of Metals, The Macmillan Company, New York, 1966, 672p.
4. S. A. Bradford, Practical Handbook of Corrosion Control in Soils, ASM International, Casti Corrosion Series, New York, 2000, 411p.
5. M. Schumacher ed., Seawater Corrosion Handbook, Park Ridge, New Jersey, 1979.
6. U. R. Evans, The Corrosion and Oxidation of Metals: Scientific Principles and Practical Applications, Edward Arnold Ltd, London, 1960.
7. W. V. Baeckmann, W. Schwenk and W. Prinz (authors and eds.), Handbook of Cathodic Protection: Theory and Practice of Electrochemical Protection Processes, Ed. by W. V. Baeckmann, W. Schwenk and W. Prinz, 3rd Ed., Gulf Publishing Co., Houston, 1997, 520p.
8. I. A. Kornfeld and V. A. Pritula, Protection of Reinforced Concrete from Corrosion by Stray Currents (in Russian), Stroyizdat, Moscow, 1964, 75p.
9. K. J. Moody, Materials Performance, Vol. 42, No. 4, 2003, 18–23.
10. I. V. Strijevsky, Underground Corrosion and Methods of Protection (in Russian), Metallurgia, Moscow, 1986, 111p.
11. C. P. Dillon, Corrosion Resistance of Stainless Steels, Marcel Dekker, Inc., Ed. by Ph.A. Schweitzer, New York, 1995, 365p.
12. R. Francis, G. Byrne and G. R. Warburton, Corrosion, Vol. 53, No. 3, 1997, 234–240.
13. N. M. Munitz, Protection of Power Cables from Corrosion (in Russian), Energoizdat, Moscow, 1982, 176p.
14. R. Yukhnevich, V. Bogdanovich, E. Valashkovsky and A. Vidukhovsky, Engineering of Corrosion Control (in Russian), Leningrad, Khimia, part 2, 1980, 224p.
15. L. Bertolini, F. Bolzoni, T. Pastore and P. Pedeferri, Stray current induced corrosion in reinforced concrete structures. In: Progress in the Understanding and Prevention of Corrosion, Ed. by J. M. Costa and A. D. Mercer, G. B. University Press, Cambridge, pp. 658–663.

16. J. A. Gonzalez, C. Andrade, P. Rodrigez, et al., Effect of corrosion current on the degradation progress of reinforced concrete structures. In: Progress in the Understanding and Prevention of Corrosion, Ed. by J. M. Costa and A. D. Mercer, G. B. University Press, Cambridge, pp. 629–633.
17. E. I. Dizenko, V. F. Novosiolov, P. I. Tugunov and V. A. Yufin, Anticorrosion Protection of Piping and Vessels (in Russian), Nedra, Moscow, 1978, 199p.
18. V. I. Glazkov, A. M. Zinevich, V. G. Kotik et al., Corrosion Protection of Extended Metallic Structures (in Russian), Handbook, Nedra, Moscow, 1969, 311p.
19. E. C. Flounders and B. M. Danilyak, Materials Performance, Vol. 34, No. 3, 1995, 17–21.
20. J. W. Walters, Stray Currents, In: Corrosion, 3rd edn, Ed. by L. L. Schreir, R. A. Jarman, and G. T. Burstein, Corrosion, Vol. 2, Corrosion Control, Butterworth, 1994.
21. I. V. Riskin, M. I. Kadraliev Yu, I. Nianiushkin, et al., Corrosion Stability of Reinforcement in Concrete under the Attack by Stray Currents, Transactions of VNIIK. Studies of Corrosion Protection of Metals in Chemical Industry, NIITEKHIM, Moscow, 1976, pp. 106–109.
22. W. V. Baeckmann, Tashchenbuch fur den Kathodischen Korrosionsschutz (in German), Vulkan–Verlag, Essen, 1987, 176p.
23. M. A. Tolstaya, I. V. Potiomkinskaya and E. I. Ioffe, Zashchita Metallov, Vol. 2, No. 1, 1966, 67–74.
24. L. M. Yakimenko, S. D. Khodkevitch, E. K. Spasskaya, et al., Platinum-titanium anodes in applied electrochemistry (in Russian), Review "Itogi Nauki i Tekhniki, Electrokhimiya" (Vol. 20). VINITI, Moscow, 1982, pp. 112–152.
25. M. A. Warne, Materials Performance, Vol. 18, No. 8, 1979, 32–38.
26. P. C. S. Hayfield and R. L. Clarke, The Electrochemical Characteristics and Uses of Magneli Phase Titanium Oxide Ceramic Electrodes, Electrochem. Society Spring Meeting, Los Angeles, 1989, 24p.

CHAPTER 3

OPERATING FEATURES OF ELECTROCHEMICAL PLANTS

Contents

3.1. GENERAL CHARACTERISTICS OF ELECTROCHEMICAL PLANTS

Electrochemical plants, as well as other chemical plants, are characterized by the high aggressiveness of reacting solutions and, in this connection, by the necessity of using equipment and pipelines made of materials of high corrosion resistance. A basic feature of the operating conditions of metallic equipment in electrochemical plants is the high probability of corrosion attack on this equipments by leakage currents coming from electrolytic baths (electrolyzers).

Despite the variety of electrochemical plants and operating conditions of the technological equipment and pipelines, they all have basic general features that allow the development of general approaches to solving problems connected with electrocorrosion and the protection of the metallic equipment in these plants.

The basic element of any electrochemical plant is the electrolyzer. The voltage of one electrolyzer or of one bipolar electrolyzer cell lies in the range of several tenths of a volt to several volts. For example, in nickel and copper electrolysis plants, the voltages of the electrolyzers are equal, respectively, to 0.2–0.4 and nearly 3 V [1, 2], and in the electrochemical synthesis of organic compounds the voltage of an electrolyzer exceeds 10 V in some cases [3].

In most cases, the central part of a large electrochemical plant consists of one, or several, electric circuits made of tens or even hundreds of electrolyzers or bipolar electrolyzer cells. A set of electrolyzers connected in series, fed from one source of current, will be hereinafter called a "set." The voltage V on the terminals of a rectifier feeding one set of electrolyzers is

$$V = vN, \tag{3.1}$$

where N is the number of electrolyzers in a set and v is the voltage of one electrolyzer.

According to expression (3.1), the potentials of electrolyzers with respect to the grounding vary linearly. On marginal electrolyzers, at positive and negative poles of the rectifier, the potentials are equal, respectively, to $V/2$ and $-V/2$. The middle of the set, where the potential is equal to zero, is called the zero point.

In modern electrochemical plants, the magnitude of the electrolysis currents reaches tens, and, in some cases, hundreds of thousand of amperes. A tendency of growth in the electrolysis current strength is observed, first of all, as a result of an increase in the current density on the electrodes. This tendency exists owing to the application of advanced materials for producing the electrodes and also to an increase their working surface area.

Most electrolysis processes are continuous. Centralized feeding of the electrolyte into the electrolytic baths and tapping of the electrolysis products are carried out through technological pipelines (headers) connected, along their full length, with each of the baths by individual tubes. The piping distribution system of an electrochemical plant includes a general header for all the technological lines and sets, and group headers for sets and groups, or blocks, of electrolyzers. Along with the electrolyzers and pipelines in electrochemical plants, there is other equipment connected through the electrolyte with headers: valves, pressure tanks, receivers, heat exchangers for heating or cooling the electrolytes and electrolysis products, pumps, etc.

There are various types of arrangement schemes of the electrolyzers by groups and by sets. In each specific case, optimal schemes are chosen [4, 5], which depend on the type of electrochemical plant, applied rectifiers and electrolyzers, etc., but which can differ even for the same type of plants. In the diaphragm electrolysis plants of Russian chlor–alkali manufactures, the electrolyzers are usually collected in sets of approximately 120 baths, based on the rated voltage of the rectifiers. At the same time, a set can consist of various numbers of groups of electrolyzers.

As an example, a typical scheme of two sets of electrolyzers is presented in Figure 3.1 for a plant for the diaphragm electrolysis of sodium salt solutions at one of the chlor–alkali manufacturers. Each of the sets includes four groups of electrolyzers, 29 electrolyzers per group. The zero point in a set is accounted for by the 58th electrolyzer. In Figure 3.1 only the brine distribution line (electrolyte feed the electrolysis process) is given.

Schemes of the distribution of technological pipelines for electrolyte feeding and for electrolysis product tapping are usually similar; they differ only in some details. A classification of schemes of the most-often-applied types of pipelines arrangement, pressure tanks and receivers inside a set of electrolyzers is given in some articles [4, 5]. Such arrangements are, in principle, common for such diverse plants as those for the electrolysis of sodium salt solutions (chlor–alkali electrolysis) and for copper electrorefining.

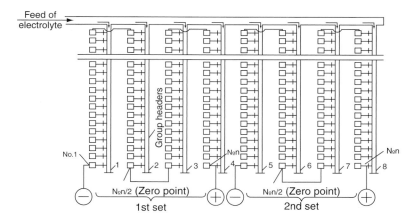

Figure 3.1   Example of a scheme of distribution of electrolyzers by groups and sets in chlor-alkali electrolysis plants.

Thus, the common feature of electrochemical plants is that for a connecting electric series of electrolyzers in a set, parallel systems exist of pipelines transporting the electrolyte to the electrolyzers and for tapping electrolysis products. With respect to a set of electrolyzers, these systems form parallel electric circuits in which a part of the industrial electrolysis current inevitably branches off, even in cases when the systems of pipelines are made entirely completely of non-conductive materials.

3.2.   LEAKAGE CURRENTS IN ELECTROCHEMICAL PLANTS

3.2.1.   Sources of leakage currents, concepts and problems connected with them

The term "leakage current" can be defined as the current lost at its source and penetrating the above-mentioned parallel circuit. In electrochemical plants, the leakage current enters the equipment and pipelines located in proximity to the electrolyzers, which are the sources of this current. This term is also used to refer to sources and consumers of current, such as high-voltage batteries [6, 7] and lines of electrified railways [8].

The pipes that connect group headers with each of the electrolyzers are always made of a non-conductive material [9]. This enables the reduction of leakage currents to pipelines and prevents short circuits between adjacent electrolyzers. As these pipes have a relatively small diameter (50–100 mm) and length (up to several meters), the selection of materials for their manufacture does not involve any difficulties. In some cases, when the destruction of several pipes occurs as a result of mechanical damage or attack by aggressive media, they can be replaced easily without an outage of other electrolyzers in the group.

The reduction of leakage current penetration to group, set and general metallic headers is achieved by the installation of insulating inserts or by electric insulation of

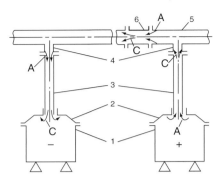

Figure 3.2 Areas of attack by leakage currents of anodic (A) and of cathodic (C) directions of elements of metallic equipment in a set of electrolyzers: 1 – electrolyzers; 2 – metallic covers; 3 – tubes of non-conductive material; 4 – metallic pipe branches (nozzles); 5 – group header; 6 – insulating insert.

adjacent pipes at their flange joints. In this case, as well as in the case of the electric sectionalization of underground pipelines, at the areas of metallic path breaks, the current leaks from the metal into the electrolyte or it leaks in from the electrolyte onto the metal (Figure 3.2).

It is important to note that, unlike in the case of stray currents, the attacked areas of the metallic structure by leakage currents and the directions of these currents, in the majority of cases, can be predetermined. This determination proceeds from the disposition and arrangement of the elements of the equipment and of the pipelines (made of conductive and non-conductive materials) with respect to the poles of the current sources. For instance, in Figure 3.2, for a given disposition of current source poles in a set of electrolyzers, the directions of leakage currents and areas of their attack are shown by the arrows. Leakage currents of the anodic (A) and of cathodic (C) directions attack the metallic headers at both sides of the insulating inserts and the metallic nozzles of the headers and electrolyzer covers.

The first data on leakage current penetration to the equipment and technological pipelines of electrochemical plants appeared in works of the 1930s [10–12]. Since the 1950s and 1960s, regular researches connected with this problem were conducted. Methods for the measurement of leakage currents were developed. These methods were used for studies of leakage current distribution along sets of electrolyzers in electrochemical plants in fields such as non-ferrous metallurgy [4, 13–15] and chemistry [5, 16–19]. The main purpose of these works was to estimate possible electric power losses in connection with leakage currents. In accordance with the problem statement and with the methods of its solution, specific works were mainly of an electro-technical character. The major paths of leakage currents coming from the electrolyzers were revealed. Along with the technological pipelines and equipment, electrolyzer and bus supports belong to such paths [4, 5, 20]. However, it was shown that the main share of the leakage currents penetrated just to the pipelines and equipment.

For example, according to Troyanovsky [13], about 80% of all leakage currents in a zinc electrolysis shop were accounted for by technological pipelines. Electric power losses in one set of this shop in the 1950s already exceeded 200,000 kW per year [13, 20].

The development and improvement of technological processes and of construction-related and technical characteristics of electrolyzer, as well as the increased volumes and capacities of electrochemical plants, required a sharp increase in the electrolysis current strength. This resulted, in turn, in an increase of leakage currents. Moreover, other things being equal, the magnitude of the leakage currents is proportional to the number of electrolyzers or cells connected in series, squared [21]. This number increases with the growth in capacity of the electrochemical plants. All these factors lead to an intensive growth of leakage currents.

The potential danger of metal corrosion by the attack of leakage currents was already indicated in early works [12, 22]. This problem has increased sharply in connection with the beginning of the wide application of titanium and other expensive and scarce metals in electrochemical plants of the chemical industry and non ferrous metallurgy. There became a situation in which the cost of corrosion damage caused by leakage currents often significantly surpassed the cost connected with energy losses.

3.2.2.   Methods of measuring and controlling leakage currents

Research of the corrosion danger by stray currents to underground and underwater pipelines and constructions consists, first of all, on the detection of their presence or absence in the construction. In electrochemical plants, where the leakage currents are irremovable, the purpose of the researches is not the detection of leakage currents, but the obtaining of quantitative characteristics of their magnitudes and their distribution.

To corrosion by stray currents are exposed, as a rule, the external surfaces of pipelines and constructions, whereas in electrochemical plants the leakage currents mainly attack the internal surfaces of pipelines and equipment. For this reason, methods of measurement and control of leakage currents in electrochemical plants essentially differ from methods that are applied for stray current measurements.

A number of experimental–design methods for the measurement of leakage current magnitudes have been based on the fact that the magnitude of the industrial electrolysis current in a set of electrolyzers is known, and it is possible to define the resistance of a parallel electric circuit to which leakage currents are flowing. For example, in work [23] a procedure has been described based on "current locations." It consists of successive current measurements after each bath or after blocks of baths. Owing to the branching of a part of a current (I_i) in a lateral circuit of the ith block, a current after this block (I_{wi}) is slightly weaker than the nominal value I_w. Thus, a current from the ith block of baths is

$$I_i = I_w - I_{wi}. \tag{3.2}$$

By means of this procedure, the presence of leakage currents was detected and their order of magnitude and even the character of their distribution were estimated in a nickel electrolysis plant. However, the leakage current I_i, calculated as the difference between the industrial currents I_w and I_{wi}, is by some orders of magnitude weaker than the industrial currents. This explains the low accuracy of such estimations.

On the other hand, the fact that it is possible to define I_i based on I_w and I_{wi}, which can attain even a thousand amperes, specifies by itself the high values of the leakage currents and indicates the high danger of their corrosion attack on metals.

An experimental-design procedure based on the principle of "voltmeter and shunt" was proposed in work [13]. This procedure enables the carrying out of more precise measurements of leakage currents from electrolyzers and definition of the character of their distribution along a set of electrolyzers. However, in these measurements, the assumption was made of the maintenance of the current symmetry with respect to zero point when a shunt is inserted into the circuit. Moreover, it was assumed that there is no interaction between the leakage currents flowing to the piping and to the ground. Therefore, the data obtained by this procedure are not sufficiently authentic.

Values of the total leakage currents flowing to the communication lines and to the ground have also been obtained in a zinc electrolysis plant [14]. These data are of interest for an assessment of energy losses, but they do not allow the assessment of the danger of corrosion attack by leakage currents to metallic pipelines and equipment.

Technological pipelines that consist of group and set headers connected by general headers of various diameters filled with electrolyte, represent, to varying degrees, an extremely complex system for electrical calculation of leakage current values. Therefore, when designing equivalent electric circuits, it is necessary to make assumptions in order to simplify the calculations. Thus, magnitudes of resistance between all electrolyzers and headers are assumed to be constant and equal. In the experimental-design procedures that were considered above, the resistance of the metallic headers was ignored.

In works [4, 5, 24, 25], analytical methods have been developed for the determination of magnitudes and the distribution character of the leakage currents using equivalent electric schemes. Also, in these cases, it was assumed that the resistances of all the pipes connecting the electrolyzers with headers were equal, and the resistances along the headers were assumed to vary linearly from one bath to another, in proportion to the length of the headers. The latter assumption implies that the diameter of the pipeline and the level of electrolyte in it were constant, and that the pipeline was made of a non-conductive material.

With the equivalent scheme constructed on the basis of the specific assumptions and represented by a number of resistors connected in series or in parallel, the magnitudes and distribution of leakage currents along a set of electrolyzers are calculated, using Ohm's and Kirchhoff's laws.

In practice, group and general headers are of various diameters and they are filled with electrolyte to various levels. Therefore, the resistance variation along the headers has a non-linear character.

It is especially difficult to obtain an equivalent electric scheme for a technological line in which a group or general headers are made partially or completely of metal. In this case, along with the Ohmic electrolyte resistances inside the pipeline sections made of non-conductive material, polarization resistances must be taken into account at the areas of metal attacked by leakage currents of anodic and cathodic directions. For example, as seen in Figure 3.2, one half of the nozzles of a metallic header are attacked by anodic currents and the other half, by cathodic currents. Areas of a metallic header that are adjacent to the ends of the insulating insert are also attacked by leakage currents of opposing directions. As anodic and cathodic characteristics of metals are not symmetric, polarization resistances of anodic and cathodic areas are not equal. Moreover, polarization characteristics vary according to the value of the external current. In the majority of cases, the polarization characteristics of metals have complex relationships, since with varying current values, the metal potential shifts to the areas where electrochemical processes of a different nature are proceeding. Owing to this, Ohm's and Kirchhoff's laws cannot be used to evaluate such systems. Refs. [21, 26] propose the use of models of electric circuits in which, along with resistors, Zener diodes are included. These diodes simulate polarizing characteristics that are similar to the Taffel dependence, which enables a more accurate estimate of the values and distribution character of leakage currents. However, the problem of producing an authentic model has still not been solved conclusively, owing to significant differences in the polarization characteristics of metals in areas of different potentials.

Alongside the analytical and experimental-design methods, instrumental methods of monitoring the leakage current were developed. Regular measurements of the magnitudes of the leakage currents at the areas of their attack on a metal structure along a set of electrolyzers gave statistical data about their distribution laws.

One of the instrumental methods used for the measurement of leakage currents is the double probe method [27]. This consists of the determination of the voltage between two identical electrodes introduced into the electrical field of the electrolyte. Various modifications of double probes have been used to measure leakage currents on pipelines. Lead electrodes were used as probes in zinc electrolysis plants [14]. In chlor-alkali plants, Ag/AgCl reference electrodes were applied as probes [28]. The double probe method is quite simple, but carrying out the measurement requires access into the pipeline or vessel, which is far from being always possible. Moreover, calibration of the probes must be carried out before each set of measurements.

These limitations are absent in a magneto-electric device developed especially for the measurement of leakage currents in pipelines and jets of electrolyte in electrochemical plants [29]. The device has a pickup in the form of one-piece ring (for measurement in jets) or a knock-down ring, consisting of two identical semi-rings (for measurement in pipelines). The rings surround the section to be measured. The pickup is supplied with a powerful Permalloy screen which decreases the effect of noise from the external electromagnetic fields, which are always present in electrolysis plants.

The double probe method allows the measurement of only the current values flowing in the electrolyte. Therefore it can be applied only for the measurement of leakage currents in pipelines made of a non-conductive material. By means of this magneto electric device, it is possible to obtain data about the total current value in a metal and electrolyte at the measured section, and it can be applied for measurements on piping of any material.

By means of specific devices, regular measurements of the leakage currents in jets of electrolyte and pipelines in a great number of electrochemical plants in the chemical industry and in non-ferrous metallurgy were carried out. Thus, the regularities of leakage current distribution along sets of electrolyzers were revealed.

One limitation of the considered device is the necessity of using a ring-shaped pickup with an internal diameter corresponding to the diameter of the pipe, for each pipe diameter at which the measurements are carried out. Moreover, the device is not able to measure leakage currents in the order of magnitude of tenths of an ampere, since the noise from the external electric fields is comparable with the measured currents. Leakage currents of such an order of magnitude are insignificant when energy losses are determined, but they are significant in the estimation of possible corrosion damage.

For estimating the danger of corrosion attack by leakage currents on metallic pipelines or on elements of equipment, it is most expedient to measure the currents directly at the areas of their path interruption in the metal structure, because it is just in these areas that currents exert corrosion attack on metal.

These measurements can be carried out by the usual electrical measuring instruments, such as a multirange millivoltammeter for direct current, supplied with a screen for protection from the influence of the electric fields from the electrolyzers [30]. On metallic piping, the measurements are carried out by contacting the instrument's terminals with the opposite sides of the metallic section at which the current interruption is executed (Figure 3.3). If there is no current interruption at this section, for example, at a flange joint, it is easy to do this by insulating the connecting bolts from the flanges by bushes and washers of insulating material.

The authenticity of the current values found by this method is verified by measurements based on the above-mentioned "voltmeter and shunt" technique. For this purpose, the measuring instrument is connected as a voltmeter and the value of the instrument input resistance is taken into account. Measurements are carried out with an inserted shunt and without a shunt, connecting the terminals of the instrument to the same section of the pipeline (see Figure 3.3).

At the pipeline section of interest, 1, where the electrolyte resistance, R_{el}, is unknown, a drop of the voltage, ΔV_0, at the insulation gasket, 2, resulting from the passage of a leakage current, I_e, is measured. After that, a shunt of a designated resistance, R_1, is introduced in parallel with the measurement instrument, 3, and a new value of the voltage drop, ΔV_1, is measured.

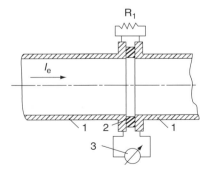

Figure 3.3 Measurement of leakage current in electrolyte at a section of current interruption by means of electrical measuring instrument (explanations are given in the text).

As the shunt resistance R_1 is negligible compared with the total electrolyte resistance in the pipe, the current leakage along the pipe does not change significantly due to the introduction of the shunt. Therefore

$$\frac{\Delta V_0}{R_{el}} \cong \frac{\Delta V_1}{R_c} = I_e, \tag{3.3}$$

where, in accordance with Kirchhoff's law, $R_c = \frac{R_1 \cdot R_{el}}{R_1 + R_{el}}$ is the total resistance at the measured section when the shunt with resistance R_1 is included. Substituting R_c into equation (3.3) solves

$$I_e = \frac{\Delta V_0}{R_1 \left(\dfrac{\Delta V_0}{\Delta V_1} - 1 \right)} \tag{3.4}$$

$$R_{el} = R_1 \left(\frac{\Delta V_0}{\Delta V_1} - 1 \right) \tag{3.5}$$

When the measurements are carried out by means of a variable resistance, it is most convenient to select the value of R_1 at which $\Delta V_1 = 0.5 \, \Delta V_0$. In this case

$$I_e = \frac{\Delta V_0}{R_1} \tag{3.6}$$

and $R_{el} = R_1$.

When leakage currents measured by an ammeter and by the "voltmeter and shunt" are close to one another (for example, within the limits of 15%), they can be considered as sufficiently authentic. The proximity of these values testifies that the influence of the metal polarization resistance and of the measuring instrument resistance is insignificant.

3.2.3. Magnitudes and distribution regularity of leakage currents along the lines of electrochemical cells

The considered methods of leakage current monitoring enabled the measurement of currents in jets of electrolyte, in branches that connect electrolyzers with group headers, and along group and general headers of technological lines of different types of electrochemical plants [4, 5, 31, 32]. The magnitudes of the leakage currents and the regularity of their distribution on pipelines made of non-conductive materials along sets of electrolyzers, were determined. The dependences of the magnitudes and distribution characters of these currents on the arrangement schemes of the electrolyzers were also studied. In all cases, the leakage currents decreased from the margin to the middle (zero point) of the electrolyzer set and current reversal took place beyond the zero point. At the same time, jumps of local currents at some branches connecting electrolyzers with group headers were observed in chlor-alkali diaphragm electrolysis plants. Current values in branches of brine lines of marginal electrolyzers in sets of these plants attained 0.5–0.8 A, and along marginal group headers the currents reached 2.2 A. From later data [33], in a set that consisted of 240 cells, leakage currents in jets reached 28 A, and along group headers they exceeded 40 A.

Leakage currents in branches of electrolyte lines feeding the baths in a copper electrolysis plant reached, in some areas, 2.8 A. However, in the majority of cases, in branches for feeding the baths and for removing the utilized electrolyte from the baths, the measured magnitudes of the leakage currents were equal to several tenths of an ampere and were in line with the rated magnitudes.

In some instances, as a result of a short circuit or of insulation failure between the electrolyzers, pipelines, equipment or ground, leakage currents can in practice reach extremely high values, greatly exceeding the designed values. Such leakage currents should be considered as accidental [24], unlike routine leakage currents, which cannot be avoided.

Especially high values of leakage currents occur in bipolar electrolyzers where the communication lines are disposed in the immediate vicinity of the electrolyzers. For example, in electrolyzers of the filter-press type used in the electrolysis of water, current leakages in communication lines attain 50 A [24]. In bipolar electrolyzers and in sets of electrolyzers connected in series, similar methods are used for the calculation of the leakage current distribution [25].

Measurements of leakage currents in metallic pipelines for technological solutions in various types of electrochemical plants, carried out in accordance with the procedure described in Ref. [30], have shown that the maximal sizes of the leakage currents in branch tubes between group headers and electrolyzers lie in the range from 0.1 A up to 1–2 A, and in group headers they can reach tens of amperes (Table 3.1).

It is seen from Table 3.1 that the maximal values of the measured leakage currents in branch tubes and in group headers of brine lines in diaphragm electrolysis plants are about 2–4 times higher than the values given in work [5]. This is explained by the facts that the data of work [5] were collected at lower values of industrial electrolysis currents and for group headers made of a non-conductive material.

Table 3.1 Measured maximal values of leakage currents (A) in electrochemical plants with metallic group headers

Type of electrolysis, medium / Objects of measurement	Electrolysis of NaCl solutions		Electro-refining of nickel, catholyte	Electrorefining of Cu, feeding of electrolyte	Electro-chemical production of perborate, electrolyte feeding
	Diaphragm electrolysis, brine	Mercury cathode electrolysis, chloranolyte			
Branches between electrolyzers and group headers	2.0	1.3	–	–	1.6
Group headers	10.0	10.0	0.8	24	–

In branch tubes connecting the electrolyzers with headers of metal and with headers of non-conductive material, the character of the leakage current distribution is similar: leakage currents reach the highest values above the marginal electrolyzers in a set, decrease towards its middle and have different directions on opposite sides of the zero point.

Measurements of leakage currents on gas lines are the most problematic. In general, the values of leakage currents on gas lines are significantly lower than on lines filled with electrolyte. However, when the humidity of the gases is high, leakage currents through the layer of condensate formed on the walls of branch tubes and of insulating inserts can be considerable, and they constitute a serious corrosion danger to metallic pipelines and equipment installed on these lines.

Analytical methods of leakage current computation on gas lines are inapplicable owing to variations in the condensate layer and, consequently, of its electric resistance inside the headers and branches. The double probe method is also inapplicable, because it is impossible to immerse the electrodes (probes) to a specific depth in the electrolyte.

Using the magneto-electric devices also does not give any positive results, since the error of these devices, amounting to tenths of an ampere, is comparable to the values of the leakage currents in the gas lines.

Measurements of leakage currents in gas lines are mostly achievable by means of the above-considered method [30]. Such measurements, outlined in Figure 3.3, were carried out along the wet chlorine lines in a chlor-alkali diaphragm electrolysis plant.

The measurements were carried out at the nozzles of the titanium group headers. Tubes for chlorine tapping from the electrolyzers (chlorine taps) made of a non-conducting material were connected to these nozzles. The measurements proved the presence of leakage currents in the nozzles that were connected to the electrolyzers by chlorine taps made of faolite (asbestos material impregnated with phenol-formaldehyde resin) or of rubber-lined carbon steel. The values of the leakage currents were significantly lower than the currents measured along the lines filled with electrolyte. Most of these values (85%) did not surpass 60 mA, but in some cases, above the marginal electrolyzers in a set, they reached 200–300 mA (Figure 3.4). Similarly to the lines filled with electrolyte, the leakage currents decreased from the end to the middle of the electrolyzer set (zero point), and past the zero point they reversed direction.

The path for leakage currents penetrating from the electrolyzers to the chlorine headers is the condensate layer that forms on the walls of the chlorine taps during the transit of wet chlorine through them. The values of the leakage currents vary widely, since they depend on the resistance of the condensate layer. This resistance, in turn, depends on the depth and composition of the condensate layer, on geometrical parameters and construction-related characteristics of the chlorine taps and on other factors that, in practice, cannot be quantitatively assessed with the required accuracy. It also depends on the working duration of the chlorine taps. If their working duration is long enough, the depth of the chlorinated layer on the walls of the chlorine taps increases. This leads to increased depths of the layer impregnated by the electrolyte (condensate) and, consequently, to a reduction of its electrical resistance.

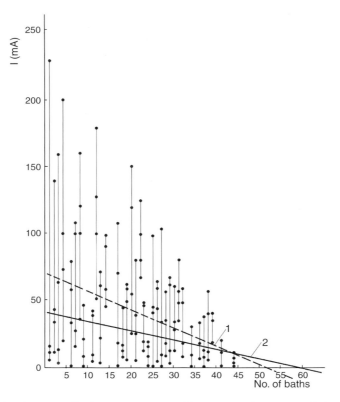

Figure 3.4 Distribution of leakage currents in chlorine taps along a set of electrolyzers and their dependence on the distance from zero point, not taking into account (1) and taking into account (2) weights of variances.

As the measured values of the leakage currents are dependant on a set of incontrollable factors, they can be considered as random values, and statistical methods can be applied for the analysis of their distribution along the header. To execute such an analysis, it was necessary, first of all, to define the distribution law of the values of the leakage currents.

As a result of measurements that were carried out at different periods of time, data were accumulated on the values of leakage currents in 35 nozzles along the titanium headers disposed at the side of the negative pole of the rectifier. Four measurements were made at each nozzle. Thus, the data for the statistical analysis consisted of 35 samples and each sample included four elements.

A hypothesis was made on a normal distribution law inside every sample. A correspondence criterion ω^2 was used for testing this hypothesis [34]. Analysis has shown a high confidence level of the assumed hypothesis which enabled treating the obtained random values by means of a least-squares technique [35]. As a result of this treatment, equations of straight lines were obtained [36], which characterize the statistical distribution of the values of the leakage currents with and without taking into account the weights of variances (Figure 3.4). It is seen that, when the weights of variances are taken into account, the zero value of the current practically coincides with the zero point in a set of electrolyzers (58th electrolyzer in a set of 116 cells). This provides support to the validity of the hypothesis on the normal distribution of the values inside a sample and indicates the applicability of statistical methods for the assessment of the values of leakage currents. On the basis of the tabulated data given in Ref. [37], it is possible to determine the confidence interval of the values of the leakage current inside a header nozzle disposed above a distinct electrolyzer in a set.

Such treatment of the results of measurements enables the application of a selective approach to materials and to means of corrosion protection that are used at different areas of technological lines along a set of electrolyzers, based on the maximal values of leakage currents at the given area. In so doing, the lifetime of construction parts made of corrosion resistant materials or that are protected by appropriate methods can be assessed at the necessary confidence level.

The above-considered statistical estimation of the distribution of leakage current magnitudes inside the nozzles of headers, provides data on the leakage currents passing through the chlorine taps. Therefore, the same data can be used to estimate the leakage currents inside the nozzles of the metallic covers of the electrolyzers, since these nozzles are connected to the opposite ends of the same chlorine taps. However, the currents in the nozzles of the covers will have an opposite direction, as can be seen from Figure 3.2.

Measurements that were carried out along group headers have shown that leakage currents inside them reached 1.3 A when chlorine taps of faolite or rubber-lined carbon steel were used [38]. This explains the intensive corrosion of the group headers at the areas of their contact with insulating inserts when these areas are not protected.

The distribution of leakage currents along chlorine taps in membrane and diaphragm electrolysis plants exhibits a similar character [39]. It is also noted in Ref. [39] that the magnitudes of the leakage currents increase evenly along the header from its margin toward the middle of the header, where they are at a maximum, and then they decrease towards the other margin of the header.

3.3.   AGGRESSIVE MEDIA

Media of electrochemical plants, like media of many other chemical and metallurgical plants, are characterized by their great variety and high aggressiveness, which is attributed to the presence of oxidizers, the high acidity or alkalinity of the solutions, and elevated temperatures and flow rates, etc. A great variety of electrolytic solutions may be found place even within the limits of one plant. For example, in the electrolysis zone of a chlor-alkali diaphragm electrolysis plant (Table 3.2), there are solutions that do not contain chlorides (hydrogen line) or that contain low (in the order of magnitude of 1 g/l) and high (at the level of saturation, e.g., wet chlorine and brine lines, respectively) chloride concentrations. The pH of the

Table 3.2   Composition and temperatures of solutions in a diaphragm electrolysis plant [41]

Technological line	Solution composition (main components)	Temperature, °C
Brine	NaCl (310 g/l)	90
Wet chlorine	Cl$_2$ (gas), water condensate (~1 g/l NaCl, pH 1.8–2)	95–98
Caustic liquor	NaOH (120 g/l); NaCl (200 g/l)	90
Hydrogen	H$_2$ (gas), water condensate	90

condensate that gets accumulates at the bottom of the wet chlorine piping is acidic (1.8–2) owing to chlorine hydrolysis. On the other hand, the caustic liquor has high concentrations of alkali and chloride.

The presence of strong oxidizers – chlorine and its oxygen compounds, permanganates, highest valence metal ions, etc., – which are formed in anodic processes, is most typical for the media of electrochemical plants.

Under the conditions of electrolysis with a mercury cathode, the hot (80°C) solution of 260 g/l NaCl saturated with chlorine, as well as the wet chlorine (like in the diaphragm electrolysis) are the most aggressive media. In electrochemical plants such as the De Nora process, piping and equipment that operate with concentrated (50%) alkali solution at a temperature of 120°C are located in the electrolysis zone [42].

Alkali solutions with concentrations of about 30% at temperatures below 100°C are typical for water electrolysis plants [17, 18] as well as for the membrane electrolysis of chloride solutions [18]. Moreover, in membrane electrolysis there are aggressive media, such as wet chlorine and chlorine-saturated chloride solutions.

Chlorine, a strong oxidizer, is the major component which defines the aggressiveness of most solutions in processes for the electrolysis of sodium salt solutions. In addition, at different stages of technological processes, these solutions may contain oxidizers such as oxygen – chlorine compounds, mainly chlorate and hypochlorite [42]. These compounds also form a part of the products which are obtained in electrolysis processes [43].

One way of producing chlorine and hydrogen is by the electrolysis of hydrochloric acid [18]. Technological solutions in this process are the most aggressive, since they are characterized by the presence of a strong oxidizer (dissolved chlorine), an activator (chlorine ion) and high acidity. The high aggressiveness of these technological media, even in the absence of the leakage currents, is the reason for the limited application of this process.

Electrochemical processes for the synthesis of oxidizers and reducers are also characterized by the great variety of aggressive media [44]. The presence of strong oxidizers (compounds of oxygen with chlorine and other halogens, perchloric and persulfuric acids, peroxides, persulfates, permanganates, etc.) in the technological solutions is most typical for this group of electrochemical plants. Technological media such as strong oxidizing acids (e.g., persulfuric acid), as well as oxidizing alkaline solutions (e.g., perborate), are present in these plants.

The aggressive media of electrochemical plants of non-ferrous metallurgy are also very varied [1–3]. A significant number of the processes in these plants are

Table 3.3 Composition and temperatures of technological media in the processes of copper and of nickel electrorefining at one of mining/metallurgical works [44, 45]

Technological media	Content of main components (g/l)					Temperature, °C
	$CuSO_4$	$NiSO_4$	$FeSO_4$	H_2SO_4	NaCl	
Cu electrorefining	130	52.5	2.7	130	8×10^{-3}	60
Ni electrorefining	–	170	–	2	57.6	85

carried out in acid, most often in sulfate solutions of different concentrations. The electrolysis of copper, nickel, cobalt, manganese and zinc employs these processes. For example, the concentrations of sulfuric acid in electrolytes of copper and nickel electrorefining at one mining/metallurgical work differ by two orders of magnitude, as seen in Table 3.3.

The technological properties of the electrolytes are adjusted by introducing into them small quantities of organic compounds that can influence the corrosion behavior of metals. For example, 50–100 g of thiourea, gelatin or joiner's glue per ton of commercial copper are added to the electrolyte of copper electrorefining.

It can be concluded, on the basis of the above data, that the presence in the technological solutions of oxidizers, which are synthesized during the anodic reactions, is typical, especially for electrochemical processes. The purpose of some electrochemical processes is obtaining a product by means of reactions of cathodic reduction [45]. However, in these cases, oxidizers that are formed in the conjugate anodic reactions are also present in the technological solutions.

The temperatures of the aqueous solutions of the electrolytes in electrochemical plants usually do not surpass 100°C, although in some cases, for example, in the above-mentioned 50% NaOH solution of the De Nora process, they can attain higher values.

REFERENCES

1. Yu. V. Baymakov and A. I. Jurin, Electrolysis in Hydrometallurgy (in Russian), Metallurgia, Moscow, 1977, 336p.
2. V. I. Beregovsky and B. B. Kistiakovsky, Metallurgy of Copper and Nickel (in Russian), Metallurgia, Moscow, 1972, 456p.
3. A. M. Sukhotin ed., Electrochemistry Handbook (in Russian), Khimiya, Leningrad, 1981, 486p.
4. N. N. Shvetzov, Tzvetnie Metalli, No. 8, 1962, 36–44.
5. N. N. Shvetzov, Khimicheskaya Promishlennost, No. 1, 1965, 60–64.
6. Sh. S. Ksenjek and I. D. Koshel, Electrokhimiya, Vol. 6, No. 10, 1587–1591.
7. B. P. Nesterov, G. A. Kamzelev, V. P. Gerasimenko and I. V. Korovin, Electrokimiya, Vol. 9, No. 8, 1154–1158.
8. A. V. Kotelnikov, V. I. Ivanov, E. P. Sedletzov and A. V. Naumov, Corrosion and Protection of Electric Railways Structures, Transport, Moscow, 1974, p.13.
9. L. M. Yakimenko, Electrolyzers with Hard Cathode (in Russian), Khimiya, Moscow, 1966, 300p.

10. A. A. Bulakh, Electrolysis of Copper (in Russian), Tzvetmetizdat, Moscow – Leningrad, 1939, 112p.
11. G. Pfleyderer, Electrolysis of Water (in Russian), ONTI – Khimteoret, Leningrad, 1935, 196p.
12. I. Ya. Sirak, Khimstroy, No. 4, 1933, 2176–2179.
13. A. V. Troyanovsky, Promishlennaya Energetika, No. 5, 1952, 12–16.
14. P. V. Sergeev, Tzvetnie Metalli, No. 6, 1954, 38–43.
15. A. S. Rumiantzev, E. P. Dubovik, M. S. Glazenap and I. T. Grigoryev, Izvestiya Vuzov. Priborostroenie, No. 3, 1958, 26–29.
16. V. B. Kogan and R. R. Ovsepian, Khimicheskaya Promishlennost, No. 8, 463–469.
17. L. M. Yakimenko, I. D. Modilevskaya and Z. A. Tkachek, Electrolysis of Water (in Russian), Khimiya, Moscow, 1970, 263p.
18. L. M. Yakimenko, Producing of Hydrogen, Oxygen Chlorine and Alkalis (in Russian), Khimia, Moscow, 1981, 279p.
19. L. M. Yakimenko, F. Z. Serebriansky and L. I. Korneev, Jurn. Prikl. Khimiyi, Vol. 44, No. 6, 1971, 1290–1297.
20. A. V. Troyanovsky, Energy Saving in Electrolysis of Zinc and Copper (in Russian), Metallurgizdat, Moscow, 1954, 161p.
21. A. T. Kuhn and J. S. Booth, J. Appl Electrochem., Vol. 10, No. 3, 1980, 233–237.
22. V. N. Poddubny, Khimstroy, No. 4, 1933, 2180–2184.
23. V. I. Juravlev and N. A. Batin, Tzvetnie Metalli, No. 1, 1953, 39–42.
24. P. P. Pirotsky and N. N. Shvetzov, Tzvetnie Metalli, No. 4, 1961, 29–34.
25. R. I. Izosenkov, V. P. Archakov, V. L. Kubasov and A. F. Mazanko, Electrokhimiya, Vol. 18, No. 9, 1982, 1255–1260.
26. M. Katz, J Electrochem. Soc., Vol. 125, No. 4, 1978, 515–520.
27. N. P. Gnusin, Jurn. Fiz. Khimiyi, Vol. 32, No. 6, 1292–1297.
28. V. V. Serkov, Khlornaya Promishlennost', No. 4, 1981, 1–4.
29. N. N. Shvetzov, P. A. Karnaushenko, M. Kh. Kulinichenko and V. A. Potapov, Electromagnetic meter of leakage currents, Author's Certificate No. 284159, Bulletin of Inventions No. 32, 1970.
30. I. V. Riskin, M. I. Kadraliev and G. P. Tutaev, Zavodskaya Laboratoriya, Vol. 40, No. 6, 1974, 705–706.
31. P. P. Pirotsky and N. N. Shvetzov, Tzvetnie Metalli, No. 8, 1960, 35–39.
32. O. P. Komlichenko, N. N. Shvetzov, V. N. Ilchenko et al., Tzvetnie Metalli, No. 10, 1975, 19–21.
33. N. N. Shvetzov, O. P. Komlichenko and Yu. M. Ivanchenko-Lirsky, Promishlennaya Energetika, No. 10, 1971, 25–28.
34. V. S. Ivanov and F. Z. Serebriansky, Gas-Oil Equipment of Generators with Hydrogen Cooling (in Russian), Energiya, Moscow, 1980, 320p.
35. N. V. Smirnov and I. V. Dunin-Barkovsky, Theory of Probability and Mathematical Statistics (in Russian), Nauka, Moscow, 1965, 511p.
36. N. R. Draper and H. Smith, Applied Regression Analysis, 3rd Edn, John Wiley & Sons Inc., 1998, 736p.
37. I. V. Riskin, A. N. Kovaliov and M. A. Kadraliev, Electrokhimia, Vol. 14, No. 2, p. 332, Deposited in VINITI, No. 3147–77.
38. V. V. Nalimov, Application of Mathematical Statistics in Substance Analysis (in Russian), Fizmatgiz, Moscow, 1960, 432p.
39. I. V. Riskin, M. I. Kadraliev and G. P. Tutaev, Corrosion Damage of Titanium Equipment an Piping in Chlor-Alkali Electrolysis Plants under the Attack by Leakage Currents. In collected articles: "Corrosion Resistant Structure Materials and their Application" MDNTP, Moscow, 1974, pp. 88–91.
40. F. Herlitz, Titanium in the Electrochemical Industry – Use and Protection, Stainless Steel World, November 2004, www.permascand.com
41. M. I. Kadraliev and I. V. Riskin, Investigation of Corrosion Protection of Steel Kh18N10T in 50% NaOH Solution at 120°C. In collected papers: "Anticorrosion Protection in Chemical Industry", MIITEKhim, Moscow, 1981, pp. 31–35.

42. V. M. Zimin, G. M. Kamarian and A. F. Mazanko, Chlorine Producing Electrolyzers (in Russian), Khimia, Moscow, 1984, 302p.
43. S. A. Zaretsky, V. N. Suchkov and V. A. Shliapnikov, Technology of Electrochemical Plants, Visshaya Shkola, Moscow, 1970, 424p.
44. M. Ya. Fioshin and M. G. Smirnova, Electrosynthesis of Oxidizers and Reducers (in Russian), Khimia, Leningrad, 1981, 212p.
45. I. V. Riskin, L. M. Lukatzky and V. A. Timonin, The Causes of Titanium Corrosion in Copper Electrolysis, Tzvetnaya Metallurgia, No. 14, 1979, 43–45.
46. I. V. Riskin, L. M. Lukatzky and V. A. Timonin, Tzvetnaya Metallurgia, No. 23, 1980, 33–35.

CHAPTER 4

USING STRUCTURES OF METALLIC MATERIALS IN ELECTROCHEMICAL PLANTS WITHOUT TAKING INTO ACCOUNT ATTACK BY LEAKAGE CURRENTS

Contents

4.1. NON-METALLIC MATERIALS

The presence of an external current does not necessitate any additional restrictions on the selection of structural non–metallic non-conductive materials if they are not intended to be used as insulating materials. Absence of porosity is one major requirement that is imposed upon insulating materials, along with the other usual requirements (high chemical resistance, strength, low cost, accessibility, etc.).

In many aggressive media, piping and equipment made of thermoplastic materials, such as PVC, polyethylene and polypropylene, are used at temperatures of up to 40–90°C (depending on the kind of material). The diameter of piping made from these materials does not surpass 200–300 mm, owing to their relatively low mechanical strength, which decreases with a rise in temperature. The tendency of these materials to undergo ageing at the surface in contact with aggressive media, especially in the presence of oxidizers, also has to be taken into account.

Teflon possesses much better thermal and chemical resistance characteristics than other thermoplastic materials. It can be applied up to temperatures of 220–250°C, and it is chemically resistant to most aggressive media, including strong acids, alkalis and oxidizers. However, its high coefficient of thermal expansion, limited molding possibilities, tendency to cold flow and relatively high price, restrict its application to the production of parts and pipes of relatively limited dimensions.

Owing to their high mechanical strength and thermal stability, glass–reinforced plastics are widely used in electrochemical plants. The required chemical resistance of these materials in aggressive media can be achieved by the selection of appropriate binders. However, in the presence of strong oxidizers, such as gaseous or dissolved chlorine, the rate of material ageing increases, which leads to a reduction in the lifetime of the piping and equipment made from glass–reinforced plastics. Reduction of the lifetime of large–size equipment made of glass–reinforced plastics is often connected with the irregular distribution of their characteristics in the mass. For example, owing to an insufficient degree of polymerization of the binder at different areas of the piping or of other kinds of structures, premature destruction of these areas takes place.

Piping and other structures made of faolite (material based on asbestos and a phenol–formaldehyde binder) are applied in many aggressive media. This material has good technological properties and a low cost [1]. It also has a satisfactory thermal stability, but its mechanical strength and, in many cases, its chemical resistance, are insufficient. The destruction rate of faolite increases in the presence of oxidizers. For example, the lifetime of piping made of faolite for hot and wet chlorine was often as low as 0.5–1 year [2]. Currently, faolite is not widely used owing to the carcinogenic properties of asbestos.

The possibilities of applying of equipment, piping and valves made of inorganic materials such as glass and porcelain in the aggressive media of electrochemical and other chemical plants [1] are limited. Owing to their fragility and low mechanical strength, these materials are used for piping of relatively small diameters (100–200 mm) and for pieces of equipments of limited size.

Of course, in electrochemical plants it is preferable to use non–conductive structural materials, which remove the danger of corrosion attack by external currents and reduce the values of leakage currents from electrolyzers. However, as seen from this short review, the possibilities of using non–metallic materials are in many cases restricted. The application of metallic materials in electrochemical plants is practically unavoidable.

4.2. TRADITIONAL METALLIC MATERIALS

Although attack by external currents is the major aggressiveness factor of the environment in electrochemical plants, the selection of structural materials in these plants is usually based, first of all, on their corrosion resistance in the given aggressive medium.

Carbon steel and stainless steel of the type 18% Cr–10% Ni (18-10) find place in electrochemical plants, as well as in other plants of the chemical and metallurgical industries, as major structural materials.

As noted in Chapter 1, in neutral media, carbon steel has an active state and corrodes, with oxygen depolarization. Its corrosion rate is quite high, especially at elevated temperatures and the high flow velocities of the aggressive media. An external anodic current attacking active steel is almost completely utilized for metal dissolution. Nevertheless, during the 1920s and at the beginning of the 1930s, carbon steel was sometimes used, without any means of protection, for brine headers in chlor–alkali diaphragm electrolysis plants [3]. This was possible owing to the relatively low leakage currents in low–power electrochemical plants. It is interesting to note that the corrosion damage in anodic areas of carbon steel headers (which was significant even in the low-power plants) was explained in Ref. [3] not by the attack of leakage currents, but as a result of iron oxidation by chlorine evolving in these areas. Currently, unprotected carbon steel is not used under such conditions.

In some cases, carbon steel was used away from the zones of leakage current action. Structural elements that were not too important, or had a great corrosion allowance and that were able to work for a long time despite significant corrosion, were made of carbon steel. For example, tube plates of carbon steel were used for brine heaters [2] in diaphragm electrolysis shops, whereas the tubes of these heaters were made of 18-10 stainless steel. The lifetime of these heat exchangers was usually about one year. Areas of contact of the expanded ends of the tubes with the tube plates were preferentially corroded. Both tube plates and tubes were damaged, which was attributed to a crevice effect. Under the attack of an external current, the lifetime of these heat exchangers went down to 6–8 months [2]. In modern practice such heat exchangers are not in use.

Carbon steel is widely used in alkaline solutions of electrochemical plants, and particularly in processes of water electrolysis [4], owing to its satisfactory corrosion resistance in alkali solutions of medium concentrations.

However, despite the fairly good corrosion resistance of carbon steel in alkaline media, it undergoes corrosion damage by external anodic currents. Rings of carbon steel for intake conduits and gas (hydrogen and oxygen) ducts of bipolar electrolyzers, and other parts, corroded under the conditions of water electrolysis. The most intensive corrosion was observed inside the ducts located above the marginal cells on the side of the negative pole of the current source. The flow of the gas–liquid mixture along these ducts was accompanied by pulsations of pressure that gave rise to pulsations of leakage currents. In the opinion of the authors of work [4], corrosion damage of the rings was connected with the appearance of the cathodic component of the pulsations. Polarization by the cathodic current leads to activation of the metal surface. During the attack by the anodic current that follows, anodic dissolution of the activated metal takes place.

Nickel is the best structural material for alkaline media that do not contain chlorides. Parts made of carbon steel with nickel coatings, including anodes for industrial electrolyzers, are applied in aqueous electrolysis media [4]. However, in accordance with the data of [4], nickel-coated parts undergo corrosion damage related to attack by leakage currents.

In processes of sodium chloride diaphragm electrolysis, carbon steel is used as a structural material for cathodic grids in electrolyzers and for piping and equipment on the lines of hydrogen and caustic liquor (solution of 120 g/l NaOH + 200 g/l NaCl) at temperatures of around 90°C [5]. Overflow devices made of carbon steel for draining caustic liquor from the electrolyzers into the header are widely used in these processes. Such devices are installed in all electrolyzers. As experience of their use has shown, the ends of these devices, from which the jet of electrolyte flows out, were severely corroded on the side of the negative pole of the power source. The most intensive corrosion occurred at the marginal electrolyzers in a set. The lifetime of the marginal devices, with wall thicknesses of 4–6 mm, usually did not surpass half a year. The dependence of the intensiveness of the corrosion of the devices on their position in the set of electrolyzers and the increase of the intensiveness of the corrosion towards the margins of the sets indicates that in this case the corrosion damage was also caused by attack by anodic leakage currents from the electrolyzer.

Inside the hydrogen headers of sodium chlor–alkali electrolysis plants, relatively small quantities of alkaline condensate are formed. The corrosion state of the headers and other equipment on the hydrogen lines does not create any serious problems, since the leakage currents along these lines are relatively weak. Carbon steel has a passive state in this environment. Some sections of the group headers sometimes undergo corrosion damage after several years of use. These sections are easily replaced during periodic routine repairs, which do not require significant expenditure or expensive materials. Therefore, any special studies of infrequent cases of metal corrosion damage were not carried out on these lines. The cases of corrosion damage were, most probably, connected to attack by leakage currents. Obtained during periodical inspections, the data on the preferential corrosion damage of sections of the group headers located above the marginal electrolyzers in a set, indicate this probability.

The headers for feeding the electrolytic baths and removing electrolyte from the baths and the valves, pressure tanks and other equipment contacting with solutions containing sulfuric acid in the copper electrorefining plants, are made of 18–10 stainless steel. Operating experience testified that the corrosion damage of this equipment depended on the location of the corroded areas with respect to the poles of the power source. Moreover, the damage was located at the terminal areas of the tubes, where jets of electrolyte flow out, and at the areas of contact of tubes, valves and other metallic parts with the parts made of a non–conductive material. As in the previous cases, these data indicate corrosion attack by the leakage currents coming from the electrolytic baths.

At the same time, corrosion of the weld seams in the piping and equipment made of 18–10 stainless steel was observed throughout the set of electrolytic baths, irrespective of the location of the corroded weld seams in the set. As is known [6], austenitic stainless steels of type 18–10 are susceptible to intercrystalline corrosion in media containing sulfuric acid in combination with oxidizers. In the case under consideration, the divalent copper plays the part of the oxidizer.

Technological piping for feeding electrolysis baths, valves and other equipment located in the proximity of the baths of the plants for the electrochemical production of sodium perborate are made of 18–10 stainless steel. Results of inspections have shown that even after more than 10 years of use, no cases of corrosion failure were noted, in spite of the high values of leakage currents along the technological lines of these plants.

Attempts to apply stainless steel 18–10 as a structural material on a line of hot caustic soda (50% NaOH, temperature 120°C) in electrolysis with a mercury cathode (De Nora process) did not give satisfactory results. Pipes with a wall thickness of 5–6 mm were perforated as a result of corrosion after twomonths of operation [7].

The damage occurred at the nozzles connected to non–metallic tubes for withdrawal of the alkali from the electrolyzers, which that were located on the side of the negative pole of the power source. Such a disposition of the damage indicates that the corrosion failure was caused by the attack of anodic leakage currents.

In the process of electrolytically producing manganese in solutions containing sulfuric acid at temperatures of 80–90°C, only high–alloyed steels and nickel-based alloys with the addition of up to 3% molybdenum and copper, possess a satisfactory corrosion resistance [8, 9]. Under these conditions, the use of steels with a lower degrees of alloying is possible only when anodic protection is provided, which maintains a stable passive state of the metals. It is clear that these steels can be applied only away from the zone of leakage current action.

Thus, the traditional structural metallic materials, carbon and stainless steels, are applied in media where they are able to maintain their passive state in the absence of attack by leakage currents. In most cases attack by anodic leakage currents is the cause of corrosion failure of piping and other structures made of these metals. Besides this, in some cases (e.g. nickel in alkaline media under the conditions of water electrolysis, stainless steel 18-10 in the electrochemical production of perborates) the metal is able to maintain its corrosion stability in the zone of leakage current action. The causes of metal stability under these conditions will be considered below.

4.3. EXPERIENCE IN THE FIELD OF TITANIUM APPLICATION

4.3.1. Preconditions for the application of titanium in electrochemical plants

The commercial use of titanium as a structural material in aggressive environments of the chemical, metallurgical and other branches of industry deserves special consideration, owing to the above–mentioned exclusive ability of titanium to maintain its corrosion stability.

In industrially developed countries, the commercial application of titanium started in the fifties of the twentieth century, when its large-scale production was mastered. Along with production growth, the use of titanium greatly increased. Currently, titanium equipment has a significant presence in different electrochemical plants, especially in the processes connected with the electrolysis of chloride solutions.

Titanium played an important part in the technical progress of electrochemical plants. The most crucial equipment at all stages of the technological processes, from preparing and feeding of the electrolyte, and up to the lines of industrial waste, was made of this material. This enabled the execution of a qualitative modernization of many kinds of equipment, improving their efficiency, reliability and other opera-tional characteristics.

It was noted in Chapter 1 that the presence of oxidizers in solutions shifts the open circuit potential of metals to the positive side, which reduces the metals passive field. As a result, traditional structural metals such as stainless steel 18-10, the corrosion stability of which is determined by its ability to maintain a passive state, undergo pitting corrosion (especially in the presence of activators) or transpassivation, even in the absence of attack by an external anodic current. At the same time, titanium, even at high concentrations of chlorides and other activators, maintains a stable passive state in such solutions, owing to the high value of its activation potential.

Media in which activators (chlorides) are present in combination with oxidizers (chlorine and its compounds with oxygen) are most typical for chlorine production processes. In this connection, titanium is the most extensively used metal for the technological lines of these processes [10, 11], which find the most widespread use in the chemical industry.

At first, titanium was used in the zones located away from the electrolysis zones of chlorine production plants. Dechlorinators (compact apparatuses which substituted bulky water dechlorination towers, made of carbon steel and lined with tiles of diabase), heat exchangers on the chlorinated water lines and waste lines containing active chlorine, as well as piping and valves, etc. were made of titanium [2, 10, 11]. Its application in these media was successful.

The universal application of titanium with electrochemically active ruthenium dioxide coating as a base for anodes was an additional incentive in the 1970s for the intensive growth of titanium application in commercial chlor-alkali processes and in the production of chlorine–oxygen compounds. Owing to their better operational parameters, these anodes substituted graphite anodes [12–14]. Later on, so-called ceramic anodes, which are based on titanium with coatings of titanium suboxides, found extensive application [15, 16].

4.3.2.   Attempts at titanium application in the zones of attack by leakage currents

During the period of intensive growth of titanium application, attempts were made to install sections of piping and other parts made of titanium in the brine lines in diaphragm chlor-alkali electrolysis plants. For example, the ends of rubber hoses for brine feed were made of titanium. The attempts to apply titanium on these lines, where the presence of leakage currents was clear, are explained by an overestimation of the corrosion stability of titanium, and by the lack of data on the corrosion and electrochemical behavior of titanium under attack by an external current. In all these cases, titanium underwent intensive corrosion of a local character. The mentioned ends of rubber hoses failed during the first four days of operation (Figure 4.1). On a chloranolyte line in an electrolysis plant with a mercury cathode, sections of group headers of titanium perforated near the insulating flanges within several days (Figure 4.2). In both of these cases, white powdery corrosion products containing titanium dioxide, which are typical for titanium corrosion under anodic polarization, were found at the corroded areas.

When the group headers for chloranolyte in the electrolysis plant with the mercury cathode were inspected, corrosion damage was also revealed at the areas of attack by leakage currents of a cathodic direction [17]. As in the case of attack by anodic currents, the damage was located close to the insulating flanges, mainly at the areas of weld seams (Figure 4.3). As opposed to the areas attacked by anodic currents, no corrosion products were found at the areas attacked by cathodic

Figure 4.1 Corrosion of hose ends of titanium in chlor-alkali diaphragm electrolysis plant under the attack of anodic leakage currents.

Figure 4.2 Perforation of group chloranolyte header of titanium in chlor-alkali electrolysis plant with mercury cathode under the attack of anodic leakage current.

currents, and the corrosion rate in these areas was significantly lower than the corrosion rate caused by anodic currents. The average time for pipe perforation to occur at a wall thickness of 3 mm was about half a year.

In nickel electrorefining plants, corrosion of titanium under attack by external anodic currents was also observed. In rake-type distributors (for catholyte distribution over the baths) located on the side of the negative pole of the power source, nozzles for connecting rubber hoses were corroded (Figure 4.4). The damages was observed at the terminal areas of the nozzles, inside them and under the hoses. Moreover, sections of the titanium group headers for catholyte feeding corroded at the areas of their contact with insulating flanges and inserts. Sometimes these areas perforated after several months of operation.

Figure 4.3 Perforation of group chloranolyte header of titanium in chlor–alkali electrolysis plant with mercury cathode under the attack of cathodic leakage current.

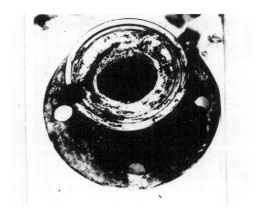

Figure 4.4 Electrocorrosion of titanium nozzle of rake-type distributor in nickel electrorefining plant.

Cathodic current attack has appeared in some areas as nickel precipitation in the form of dendrites.

Under attack by an external cathodic current, titanium hydrogenation can occur along with corrosion. Data on the hydrogenation of titanium cathode blanks [18] which takes place in nickel electrorefining plants point to such a possibility.

Titanium piping and equipment in copper electrorefining plants are also susceptible to corrosion by external anodic currents. As in other kinds of electrochemical plants, corrosion damage was noted in the areas of metallic pipes joined to the pipes of non-conductive materials and at the terminal areas of the pipes where jets of electrolyte were flowing out. In particular, titanium valves installed on polyethylene piping and at terminal areas of titanium pipes for feeding the electrolyte to pressure tanks were corroded.

Welded bottoms and nozzles of degassing tanks were among other titanium equipment attacked by anodic leakage currents in these plants. The nozzles were connected with rake-type polyethylene distributors for feeding electrolyte into the baths. Only weld seams were corroded at their bottoms. The tanks were positioned directly on the baths or on the bottom polyethylene plates. The walls of the baths were made of bricks that were impregnated with electrolyte, and the bottom plates were always damp due to electrolyte spillage, which led to the appearance of leakage currents. The location of the weld seams on the tank bottom coincided with the zone of leakage current attack.

As under the conditions of nickel electrorefining, copper precipitation in the form of dendrites took place in the copper electrorefining plant.

Consideration of the results of the inspections in the electrochemical plants pointed to the conclusion that in the absence of means of protection against corrosion, titanium, like other structural metallic materials, is susceptible to corrosion attack by leakage currents.

4.3.3. Preconditions and experience of titanium application in wet chlorine lines of electrochemical plants

As was already noted, most of the non-metallic structural materials undergo chlorination in the environment of hot, wet chlorine. This leads to the destruction of structures made of these materials during relatively short periods of their operation. Titanium is the only metallic structural material that is able, in the absence of attack by leakage currents, to maintain its corrosion stability in wet chlorine and in chlorinated water almost for ever. Owing to this, titanium found extensive application, first of all, in wet chlorine headers. It was suggested that leakage currents from electrolyzers do not penetrate to the gas lines.

A header for wet chlorine in a modern sodium chloride electrolysis plant comprises a branched system of pipes with diameters of from 200–300 mm to 900–1200 mm (group and general headers) and of lengths in the order of 1000 m. Frequent replacement of this construction, which requires a lot of materials along with direct material and labor expenditures, leads to long periods of stopages in the operation for overhauls. Therefore, the application of titanium headers was estimated in the 1970s to be an important step in the technical progress of the chlorine producing industry. Sometime later, titanium covers for diaphragm electrolyzers began to be manufactured, since the covers made of non-metallic materials, like the headers, were destroyed by the hot, wet chlorine.

During the same period of time, signs appeared of cases of titanium corrosion failures in chlor-alkali electrolysis plants.

The first data on the corrosion of titanium in wet chlorine were published by American researchers Sheppard and Gegner [19] and Gleekman [20]. In Ref. [19], intensive corrosion was noted of titanium branch tubes (chlorine taps) in the crevices under the sealing at the areas of the connection of these tubes to the branches of the group headers and the covers of the electrolyzers. Gleekman [20] also described corrosion damage to chlorine taps made of titanium in crevices and, what is more, corrosion damage to sections of titanium headers in crevices, under the sealing of the piping joints. These damages were observed under the operating conditions. It was noted in Refs. [19, 20] that titanium dioxide was the major corrosion product in the corroded areas. At the same time, it was indicated that while one of the ends of a pipe section was intensively corroded, the other end, which was also inside a crevice (under the sealing), showed no signs of corrosion. The cases of corrosion resistance of titanium in the bulk of the chlorinated solution and of its corrosion failure in the same solution inside crevices was determined by the author of the article [20] as abnormal. The corrosion failure was explained to be a result of a crevice effect. Based on the work by Bomberger [21], a mechanism for the development of crevice corrosion in wet chlorine and chlorinated solutions was proposed. The author suggested that in the first stage of corrosion development in deep crevices, the chlorine loses water, which is spent on the initial corrosion process. This leads to intensive corrosion of the titanium in the dewatered chlorine. The fact that one end was, and the other end was not, damaged inside the crevices under the sealing, was explained by the author as resulting from the low reproducibility of the obtained data, owing to the differing geometric parameters of the crevices.

Although all the considered cases of corrosion were revealed at the branches and headers of titanium located in proximity to the electrolyzers, the above-mentioned works did not even refer to the possibility of corrosion attack of titanium by leakage currents. Apparently, the authors did not believe that the piping that transports gas and not electrolyte could be penetrated by leakage currents. However, in 1967 the same authors suggested that under such conditions, attack of titanium by external currents is possible [22].

Extensive inspection of electrochemical plants that was undertaken in the Soviet Union during the 1980s confirmed the corrosion attack of titanium by leakage currents on the wet chlorine lines. The damage was mainly found inside crevices under the gaskets of flange connections. At first glance, this corroborated the conclusions of the American researchers. However, careful analysis has shown that the damage occurred only in the piping sections that, according to the scheme of the electric circuits, were attacked by leakage currents of an anodic direction. In particular, the flanges of the titanium group headers on the side of the negative pole of the power source corroded, like in the above-mentioned brine lines. The most intensive corrosion was observed above the marginal electrolyzer in a set. Thus, the results of the inspection unambiguously pointed to the corrosion attack of titanium by anodic leakage currents.

Moreover, the corrosion damage was found not only inside the crevices under the gaskets, but also in areas adjacent to flanges, and at a significant (about 100 mm) distance from the flanges, where the crevice effect cannot appear. An example of corrosion damage of such a character, at a terminal area of a section of a titanium group header that was connected to an insulating insert, is shown in Figure 4.5.

At the branches of the titanium group header that were disposed on the positive side, with respect to the poles of the power source, no corrosion damage was found. Also, no signs of corrosion were found under the gaskets of the titanium flange connections of adjacent sections of titaniumpipes that were electrically connected through connection bolts. So, in both these cases, the crevice effect was absent.

Further inspection of the titanium covers for electrolyzers have shown that only the nozzles for connecting the chlorine tap tubes of insulating material to those

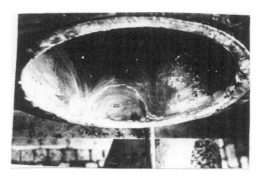

Figure 4.5 Corrosion of a section of a group titanium header near its contact with an insulating insert.

covers that were disposed on the positive side of the power source were corroded. On the opposite side of the zero point, such nozzles were not corroded. Thus, corrosion of these nozzles also can be explained only by an attack of leakage currents of an anodic direction.

Corrosion products containing titanium dioxide, which is formed as a result of attack by anodic currents, were found at the corroded areas.

4.3.4. Laboratory and industrial tests of titanium resistance against crevice corrosion in wet chlorine and in chlorine-saturated water: Comparison of results

To estimate whether titanium is prone to crevice corrosion in wet chlorine and in chlorinated aqueous solutions, corrosion tests were carried out on titanium specimens with artificial crevices in a saturated chlorine solution of 100 g/l NaCl, at temperature of 95°C.

The testing procedure is described in Ref. [23]. Plate-shaped rectangular specimens with the dimensions $25 \times 50 \times 2.5$ mm were placed in a yoke of Teflon. Crevices, of 0.1 mm width between each pair of specimens, were set up with the help of narrow (2 mm width) ribbons of Teflon foil, placed at the opposite long edges of the specimens. The specimens were pressed together using Teflon screws. Each pair of specimens was insulated from adjacent pairs by fluorine rubber gaskets. The yokes were placed inside the cells so that the working surfaces of the tested specimens were disposed vertically. For simultaneous testing in a chlorinated solution and in wet chlorine, one part of the specimens was tested under complete immersion into the solution, and another part was placed above the solution. The duration of the tests was 500 h.

In none of the tested specimens were signs of crevice corrosion found, regardless of location of the specimens inside the cells.

Under the working conditions, tests were carried out on specimens with artificial crevices in accordance with the procedure described in Ref. [24]. In this case, pairs of square-shaped titanium specimens of dimensions 25×25 mm were tested. Artificial crevices of 0.1 mm width were also set up with the help of narrow ribbons of Teflon foils placed at the opposite edges of each pair of specimens. Several pairs of specimens were mounted in a set with the help of Teflon bolts which passed through the central orifices of the specimens. These sets were placed inside electrolyzers, under the covers. In six month long test, also no signs of crevice corrosion were found.

Along with the testing of specimens with artificial crevices, industrial corrosion tests were carried out on titanium branch tubes for chlorine withdrawal (chlorine taps) from electrolyzers to group headers made of faolite. Furthermore, chlorine taps made of titanium were tested, with short pipes (100 mm long) of Ti alloy with 0.2% Pd welded to the terminal part of the chlorine taps. The terminal parts of the chlorine taps were embedded into electrolyzers' covers of concrete. When the chlorine tap with the short pipe of Ti alloy–0.2% Pd was tested, the pipe of the alloy was completely embedded inside the cover, together with the adjacent terminal area of the titanium chlorine tap.

The tests showed that the areas of the chlorine taps embedded into the covers, whether of titanium or of Ti alloy–0.2% Pd, had undergone corrosion damage when they were located on the side of the positive pole of the power source (Figures 4.6 and 4.7). The titanium pipes were perforated, and the short Ti–0.2% Pd pipes were also badly corroded, whereas this alloy usually possesses high resistance against crevice corrosion even in a 5% solution of HCl at 80 C [23]. The opposite terminals of the chlorine taps (joined to the faolite headers) were corroded on the side of the negative pole of the power source.

The results of the laboratory and industrial tests ultimately confirmed that under the considered conditions, titanium is not susceptible to crevice corrosion in wet chlorine and in chlorinated solutions. The only cause of the corrosion of the chlorine taps made of titanium is the attack by anodic leakage currents coming from the electrolyzers.

The areas of attack by leakage currents coincided with the areas where the appearance of crevices between the chlorine tap and the cover of the electrolyzer

Figure 4.6 Corrosion of a chlorine tap made of titanium at the area of its embedding inside a cover of concrete.

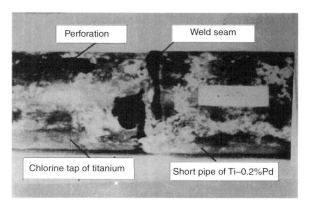

Figure 4.7 Corrosion of a short pipe of Ti alloy-0.2% Pd and perforation of the area adjacent to it of the terminal part of the titanium chlorine tap embedded inside a cover of concrete.

were possible. It is probable that this was the reason for the authors of Refs. [19, 20] to point to the crevice effect as the cause of the titanium corrosion in wet chlorine and in chlorinated solutions.

The lower corrosion rate of the Ti alloy-2% Pd compared with titanium, which was observed at the chlorine taps, may be explained by the presence of palladium in the alloy. When the activation potential, E_a, of the alloy is attained under the attack of an external anodic current, enrichment of the corroding metal surface with palladium occurs. A similar effect takes place under active dissolution of this alloy in acids [25, 26]. Palladium which was accumulated on the alloy surface, is an anodically active metal, and some share of the external current is spent on the palladium particles for the evolution of chlorine. As a result, the share of current which is spent for titanium dissolution, is reduced.

Thus, on all electrochemical plant lines, including gas lines, titanium can only be applied without some means of protection against corrosion, in places that are distanced from zones of attack by leakage currents .

REFERENCES

1. I. Ya. Klinov, P. G. Udima, A. V. Molokanov and A. V. Goriainova, Corrosion Resistant Chemical Equipment, Handbook (in Russian), Mashinostroenie, Moscow, 1970, 591p.
2. A. M. Sukhotin and A. L. Labutin eds, Corrosion and Protection of Chemical Equipment, Handbook (in Russian), Vol. 6, Khimiya, Leningrad, 1972, 373p.
3. V. N. Poddubny, Khimstroy, 1933, No. 4, 2180–2184.
4. L. M. Yakimenko, I. D. Modilevskaya and Z. A. Tkachek, Electrolysis of Water (in Russian), Khimiya, Moscow, 1970, 263p.
5. L. M. Yakimenko, Electrolyzers with Hard Cathode (in Russian), Khimiya, Moscow, 1966, 300p.
6. V. Cihal, Intercrystalline Corrosion of Stainless Steels (in Russian), Khimiya, Leningrad, 1969, 231p.
7. M. I. Kadraliev and I. V. Riskin, Investigation of Corrosion Protection of Steel Kh18N10T in 50% NaOH Solution at 120 C. In collected papers: "Anticorrosion Protection in Chemical Industry", MIITEKhim, Moscow, 1981, pp. 31–35.

8. Sh. L. Gelagutashvili, S. N. Manjgaladze, V. A. Makarov et al., Zashchita Metallov, Vol. 18, No. 6, 1982, 915–917.

9. Sh. L. Gelagutashvili, S. N. Manjgaladze, V. A. Makarov et al., Zashchita Metallov, Vol. 15, No. 1, 73–75.

10. Yu. S. Ruskol, Titanium Structural Alloys in Chemical Plants (in Russian), Khimia, Moscow, 1989, 286p.

11. A. Ullman, Cost saving in chlorine plants by benefiting from the unique properties of titanium (Permascand AB), Modern Chlor-Alkali Technology, Royal Soc. of Chemistry, 2001, Vol. 8, This is handbook Ed, by J. Moorhouse. 282–294.

12. L. M. Yakimenko, Electrode Materials in Applied Electrochemistry (in Russian), Khimiya, Moscow, 1977, 264p.

13. V. N. Antonov, V. I. Bistrov, V. V. Avksentiev et al., Khimicheskaya Promishlennost, 1974, No. 8, 600–603.

14. A. F. Mazanko, V. L. Kubasov and V. B. Busse-Machukas, Application of metal-oxide anodes in producing of chlorine, alkali and oxygen-chlorine compounds. Proceedings of Fourth All-Union Seminar "Dimensionally Stable Anodes and Their Application in Electrochemical Processes", NIITEKhim, Moscow, 1979, pp. 3–5.

15. H. Boder, H. Herbst and O. Rubisch, Chem. Eng. Technik, Vol. 49, No. 4, 1977, 331–333.

16. M. O. Coulter ed., Modern Chlor-Alkali Technology, Ellis Horwood Ltd, Chichester, Section B, Coatings for Metal Anodes and Cathodes, 1980, pp. 99–149.

17. I. V. Riskin and M. I. Kadraliev, Titanium Protection from Electrocorrosion on Lines of Brine and Chloranolyte of Chlor-Alkali Plants. In collected articles: "Investigations of Metals Protection against Corrosion in Chemical Plants", NIITEKhim, Moscow, 1978, pp. 50–55.

18. A. I. Levin, Electrochemistry of Non-Ferrous Metals (in Russian), Metallurgia, Moscow, 1982, 256p.

19. R. S. Sheppard and P. J. Gegner, Chem. Eng. Prog., 1965, Vol. 4, No. 3, 114–117.

20. L. W. Gleekman, The abnormal corrosion behavior of titanium in wet chlorine gas and in chlorine solution. Proceedings of Third International Congress on Metals Corrosion, Vol. 2, 1969, Moscow, pp. 352–366.

21. H. B. Bomberger, Indu. Eng. Chem., Vol. 56, No. 8, 1964, 55–58.

22. L. W. Gleekman, P. J. Gegner, E. J. Bohlmann et al., Materials Protection, 1967, Vol. 6, No. 10, 22–27..

23. I. V. Riskin, Z. I. Ladojina and N. D. Tomashov, Zashchita Metallov, Vol. 8, No. 2, 1972, 177–181.

24. K. Shiobara and S. Marioka, J. Japan Inst. Metals, 1971, Vol. 35, No. 10, 980–984.

25. N. D. Tomashov, M. N. Shulepnikov and Yu. I. Ivanov, Zashchita Metallov, Vol. 1, No. 1, 1965, 122–133.

26. A. A. Uzbekov, I. V. Riskin, Z. I. Ladojina and N. D. Tomashov, Zashchita Metallov, Vol. 8, No. 1, 8–14.

CHAPTER 5

CORROSION BEHAVIOR INVESTIGATIONS OF TRADITIONAL STRUCTURAL METALLIC MATERIALS IN ELECTROCHEMICAL PLANT MEDIA, TAKING INTO ACCOUNT ATTACK BY LEAKAGE CURRENTS

Contents

5.1. CARBON STEEL IN NEUTRAL, ALKALINE AND CHLORIDE-ALKALI MEDIA

5.1.1. Neutral solutions of salts

The corrosion and electrochemical behavior of iron and carbon steel in neutral solutions of salts such as chlorides and sulfates is well studied [1]. With access to air, carbon steel actively corrodes in distilled water and in aqueous solutions of these salts. Nevertheless, as was noted in the previous chapter, in the early stages of the development of electrochemical plants, carbon steel equipment was used in the zones of action of leakage currents coming from the electrolyzers. In these cases, the external anodic current, which attacked the actively corroding metal, was spent almost completely on its dissolution, in accordance with Faraday's laws. Therefore, using carbon steel in neutral solutions in electrochemical plants under conditions of attack by leakage currents is inadmissible without complete isolation of the metallic structures from contact with an aggressive environments.

5.1.2.   Alkaline solutions

As seen from the Pourbaix diagram (Figure 1.1), iron and carbon steel are in a passive state in dilute alkaline solutions and, consequently, they maintain their corrosion stability in the absence of attack by external anodic currents [2, 3]. Iron passivation in these media is connected with formation of oxide compounds on the metal surface. The depth of the layer formed by these compounds may be in the order of magnitude of one monolayer, and the passive state can be reached when only a part of the metal surface is covered by the oxide [4].

During the anodic polarization of passive iron in alkali solutions, different types of oxide and hydroxide compounds, including Fe^{2+} and Fe^{3+}, ions are formed. The compositions of these compounds depend on the potentials at which the corresponding reactions are proceeding and on the concentrations and temperatures of the solutions.

The reaction of anodic oxygen evolution

$$2OH^- \rightarrow \frac{1}{2}O_2 + H_2O + 2e^- \qquad (5.1)$$

is thermodynamically possible in 1 N alkali solution at potentials that are more positive than 0.4 V [5]. The overvoltage of the oxygen evolution on iron at low current densities is 0.25 V [6]. Therefore, at potentials that are more positive than ~0.65 V in 1 N alkali solution (and in more concentrated solutions at more negative potentials), the major share of the current will be spent on oxygen evolution. In accordance with the potential–pH diagram obtained for hot alkaline solutions [7], the existence of lepidocrocite (γ-FeOOH) and of magnetite (Fe_3O_4) on the metal surface is possible in this range of potentials.

$$Fe_3O_4 + 2H_2O = 3\gamma\text{-FeOOH} + H^+ + e^-, \qquad (5.2)$$

and the Fe_3O_4 layer has a friable structure.

It was of interest to estimate to what area the potential of the carbon steel will, most probably, shift under the action of an anodic current, and to consider the corrosion resistance of the carbon steel in this area.

The results of potentiostatic investigations of carbon steel in 1 N NaOH solution at a temperature of 90°C and at a potential scanning rate of 0.72 V/h (Figure 5.1, curve 1) are in good corroboration with the data in Ref. [8]. At potentials from stationary (−0.1 to 0 V), and up to the potential of the start of oxygen evolution (0.7 V), the steel maintained its passive state. In the area of oxygen evolution, the curve had a Taffel dependence. The current density on the passive steel was only $(1–2) \times 10^{-5}$ A/cm^2. Therefore, it could be expected that even on small values of current density, the potential will shift to the area of oxygen evolution. This was confirmed in experiments that were carried out under conditions of galvanostatic polarization. The potential shifted to the value 0.73–0.75 V at a current density of 1.5 mA/cm^2. Tests lasting five hours have shown that the weight losses at this current density are insignificant, less than 0.1 g/m^2 h.

At higher current densities, the potentials and the weight losses increased significantly. For example, in a 5 h long experiment at current densities of 20 and 50 mA/cm^2, the potentials stabilized at the values 0.80–0.82 V and 0.83–0.87 V,

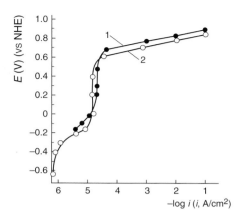

Figure 5.1 Anodic polarization plots on carbon steel St 3 in solutions: 1 – 1N NaOH; 2 – 120 g/l NaOH + 200 g/l NaCl. Temperature 90°C.

respectively. The weight losses of carbon steel therewith were, respectively, 0.68 and 1.1 g/m^2 h, or, in current density units, 0.5 and 0.8 mA/cm^2. The shares of the current that were spent on iron dissolution, if the transfer of the Fe^{3+} ions into solution is accounted for, were, at 20 and 50 mA/cm^2 respectively, 2.5 and 1.6%. At 50 mA/cm^2, the share of the current spent on iron dissolution was lower than at 20 mA/cm^2, but the absolute value of the weight loss at 50 mA/cm^2 was higher.

The corrosion of carbon steel in the area of oxygen evolution can be attributed to the formation of soluble compounds such as FeO$_4^{2-}$ at high anodic potentials [9].

An increase in the corrosion rate in the area of anodic oxygen evolution was observed by many researchers [2, 10]. The stage of adsorption of oxygen contained in the water molecule, onto the metal surface precedes the process of oxygen evolution in the form of bubbles. Following this, further oxygen transfer from the anodically dissolving "cells" of iron oxide to the adsorbed layer of oxygen (which plays the part of a buffer capacity) would be facilitated [11]. This is considered to be one of the causes of the growth in the corrosion rate under the conditions of oxygen evolution. The possibility of mechanical destruction of the friable surface layer of magnetite by the bubbles of evolving oxygen also has to be taken into account.

In accordance with the existing data [6], the action of external cathodic currents does not lead to the corrosion of carbon steel in alkaline media.

5.1.3. Chloride-alkali solutions

When alkaline media contain chlor–ions, local corrosion of carbon steel can develop even in the absence of attack by external anodic currents. For example, pitting corrosion of carbon steel was observed in alkali solution (pH 11) that contained only 0.5 mg/l of NaCl [12]. Alkalization of water under conditions of air access to the water is considered to be a dangerous means of corrosion protection for steam power equipment, because of the probability of the development of pitting corrosion [13].

The activating action of chlor-ions in the anodic polarization of iron and the local character of corrosion were first revealed in alkaline solutions [14]. An increase in the alkali concentration suppresses the activating action of chlor-ions [9, 15]. On the other hand, an increase in the chloride concentration, as well as an elevation of the temperature, facilitates the metal activation, which can occur at potential values that are more negative than the potential of oxygen evolution. Thus, the value of the activation potential of the metal, E_a, in chloride-alkali solutions, depends on ratio of the concentration of chloride and hydroxyl ions and on the temperature.

The mechanism of metal activation in solutions containing ion activators, considered in Chapter 1, is applicable for alkaline solutions, since in both cases the activation takes place as a result of substitution of the passivator (oxygen) by the adsorbed chlor-ion.

According to the data [16], carbon steel maintains its corrosion stability up to temperatures of 140°C in a solution of 120 g/l NaOH + 200 g/l NaCl. Such a solution is typical in chlor-alkali diaphragm electrolysis plants. It was of interest to estimate the carbon steel corrosion stability in this solution in the presence, and in the absence, of the action of external anodic currents.

Corrosion tests have shown that the rate of corrosion of carbon steel St 3 in this solution at a temperature of 90°C, in the absence of an external current, was less than $0.1 \text{ g/m}^2 \text{ h}$. The open circuit potential of carbon steel under these conditions was -0.65 to -0.75 V. This value is much more negative than the potential value in 1 N NaOH solution (see Figure 5.1, curve 1) and, as seen from Figure 5.1 (curve 2), carbon steel had a passive state at this potential, which explains its corrosion stability.

An anodic current growth, connected with oxygen evolution, was observed in a chloride-alkali solution at potentials above 0.5 V (see Figure 5.1, curve 2). Along with oxygen evolution, a considerable share of the current was spent in this area on metal dissolution. At a current density of 20 mA/cm^2, the stabilized potential value was equal to 0.7 V, and the corrosion rate at this potential reached $4 \text{ g/m}^2 \text{h}$, or, in current density units, 3 mA/cm^2. The current share that was spent on iron dissolution, equal to 15%, was significantly higher than that in 1 N NaOH solution.

Thus, in the absence of an external current, carbon steel maintains its corrosion stability in chloride-alkali, as well as in alkali solutions. Under attack by external anodic currents, the potential of carbon steel shifts to the area of oxygen evolution, where the corrosion rate of carbon steel in alkaline solution becomes significant. In chloride-alkali solution the corrosion rate of carbon steel further increases.

5.2.   STAINLESS STEEL 18-10 IN ALKALINE AND ACID MEDIA

5.2.1.   Solution of 50% NaOH at 120°C; comparison of corrosion behavior of stainless steel 18-10 and of nickel

The open circuit potential of stainless steel 18-10 in a concentrated solution of NaOH, which is equal to -0.9 V, refers to the area of active iron dissolution [7]. Apparently, this is the cause of a noticeable – in the order of 0.3 mm/year – rate of corrosion of steel 18-10 in this solution, in the absence of attack by an external current, which is in agreement with the data given in Ref. [17].

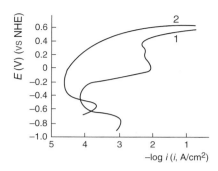

Figure 5.2 Anodic polarization plots on stainless steel 18-10 (1) and on nickel NP-2 (2) in a solution of 50% NaOH at a temperature of 120°C.

As already noted in Section 4.2, in electrolysis plants with the mercury cathode, under conditions of attack by leakage currents of an anodic direction, pipes made of stainless steel 18-10 in the hot caustic soda lines corrode at a much higher rate.

Studies of stainless steel 18-10 and of nickel NP-2 that were carried out in this connection revealed peaks on the potentiodynamic polarization curves, which are related to the above-mentioned active dissolution of the metal (Figure 5.2). For steel 18-10, the peak appears at potential -0.9 V; for nickel, the peak is located at a more positive potential, -0.6 V. In Ref. [18], it is also noted that an area of active nickel dissolution exists in media of similar compositions.

The current density growth on steel 18-10 at potentials (-0.3) to (-0.1) V is explained in Ref. [9] by the transpassivation of chrome, which is followed by the formation of soluble hexavalent chrome compounds.

$$Cr(OH)_3 + 5OH^- = CrO_4^{2-} + 4H_2O + 3e^- \tag{5.3}$$

$$CrO_3^{3-} + 2OH^- = CrO_4^{2-} + H_2O + 3e^-. \tag{5.4}$$

The equilibrium potentials of these compounds in alkaline media are equal, respectively, to -0.13 and -0.19 V.

The passivation of steel 18-10 at more positive potentials is explained in Ref. [19] by the formation of the oxide Ni_2O_3. At high anodic potentials, iron passes into solution in the form of FeO_4^{2-} ions [20].

The Taffel area of oxygen evolution on nickel is disposed above the potential of 0.4 V (Figure 5.2). The high corrosion stability of nickel in alkaline solutions is well-known. Its corrosion rate in 50% NaOH solution, under evaporation conditions, is 0.06 g/m^2 day [3].

It was found in Ref. [21], that under anodic polarization, nickel remains passive and the stability of its passive state rises with a rising anodic current, which is stipulated by the highly protective properties of passivating oxides forming on the nickel surface [19]. During anodic polarization, ions of nickel practically do not participate in the anodic process.

At the same time, there are indications of delayed nickel passivation in alkali at temperatures above 120°C and on active dissolution of the metal under these conditions [6]. Therefore, it was of interest to consider the anodic behavior of nickel, and to estimate its corrosion stability compared with the behavior of steel 18-10.

It is seen from the anodic polarization curve 2 (Figure 5.2) that, in accordance with the data [18], no transpassivation area exists on nickel. In the range of potentials from −0.2 to 0.3 V, the current density on nickel is lower than that on steel 18-10 by more than two orders of magnitude. Oxygen begins to evolve on nickel at a potential of about 0.3 V, which is 0.1 V more negative than on steel. Nevertheless, the Taffel branch relating to oxygen evolution is disposed on nickel at a potential 0.1 V more positive than on steel 18-10, and the current densities on this branch are an order of magnitude lower than those on steel. The share of current that is spent on the anodic dissolution of steel components in the area of oxygen evolution does not exceed a few percent of the total anodic current. Therefore, a comparison of curves 1 and 2 of Figure 5.2 points to the conclusion that the overvoltage of oxygen evolution on nickel is higher than that on steel 18-10. This can be explained by the active dissolution of steel components in the trans-passivation state, which increases the electrochemical activity on the steel surface.

Galvanostatic studies at current densities from 1 to $100 \, \text{mA/cm}^2$ have shown that the weight losses of nickel are insignificant. The potential of nickel shifted to the potentials of oxygen evolution at these values of current densities.

To estimate the corrosion rate of steel 18-10 at potentials related to the areas of transpassivation and oxygen evolution, 5 h long corrosion tests were carried out under potentiostatic conditions, at potentials, −0.11 and 0.45 V, respectively. It was shown that the corrosion rate at the transpassivation potential (−0.11 V) is higher, than that at the potential of oxygen evolution: 8 and $2 \, \text{g/m}^2 \, \text{h}$, respectively, although the current density at −0.11 V is about two orders of magnitude lower than that at 0.45 V. This can be attributed to the secondary passivation of stainless steel in the area of oxygen evolution.

These results accord with the results of the 5 h long corrosion tests obtained under conditions of galvanostatic polarization. At a current density of $1 \, \text{mA/cm}^2$, the potential of steel 18-10 fell in the area of chrome transpassivation (Figure 5.3, curve 1), and its corrosion rate was the highest (Table 5.1). At a current density of $5 \, \text{mA/cm}^2$ the potential during the first 2 h remained in the area of transpassivation and then it drastically shifted to the area of oxygen evolution (curve 2). The higher the current density, the faster the metal potential shifted to the area of oxygen evolution. As the current density was increased from 5 to $50 \, \text{mA/cm}^2$, the corrosion rate of steel 18-10 decreased (curves 3−5). However, a further rise in the current density to $100 \, \text{mA/cm}^2$ led to a rise in the corrosion rate.

Thus, the corrosion stability of stainless steel in concentrated alkali solutions is quite low, even in the absence of attack by external anodic currents. The stability decreases still further under the action of both low and high external anodic currents at which steel undergoes corrosion at the potentials of either transpassivation or of oxygen evolution. At the same time, nickel maintains its corrosion stability in a large range of potentials and current densities, owing to its stable passive state.

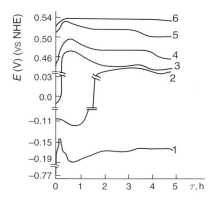

Figure 5.3 Chronograms of potentials at galvanostatic polarization of stainless steel 18-10 in a solution of 50% NaOH at temperature 120°C at current densities, (mA/cm^2): 1 – 1; 2 – 5; 3 – 10; 4 – 20; 5 – 50; 6 – 100.

Table 5.1 Corrosion rate of stainless steel 18-10 in a solution of 50% NaOH at 120°C after 5 h galvanostatic testing at different current densities

Current density, mA/cm^2	1	5	10	20	50	100
Corrosion rate, g/m^2 h	6.3	5.2	2.4	1.6	1.0	5.5

Under the action of an external anodic current, the electrochemical process of oxygen evolution takes place on the passive nickel surface.

5.2.2. Technological solution for producing sodium perborate

The technological solution for producing sodium perborate, which contains 150 g/l Na$_2$CO$_3$, 25 g/l NaHCO$_3$ and 7 g/l B$_2$O$_3$ at a temperature of 15°C, is alkaline (pH 10).

During the 5 h long corrosion tests of stainless steel 18-10 specimens, no noticeable weight losses of the specimens were revealed, and the metal surface remained shiny.

Studies under the conditions of potentiodynamic polarization have shown that the open circuit potential of steel (about 0 V) relates to its passive state. On the anodic polarization curve, at a potential above 1.05 V, there is a bend corresponding to the beginning of oxygen evolution (Figure 5.4), which has a Taffel character.

At anodic galvanostatic polarization by a current density of 20 mA/cm^2, a stable potential of 1.35 V was immediately achieved. The corrosion rate at this potential value did not exceed 0.06 g/m^2 h.

Under the conditions of potentiostatic polarization at a potential of 1.8 V, the corrosion rate was insignificant, below 0.1 mA/cm^2. Only at a potential value as high as 1.9 V, at which the current density reached a value of 100 mA/cm^2, did the corrosion rate of the steel increase to 0.7–1.4 g/m^2 h. This growth in the corrosion

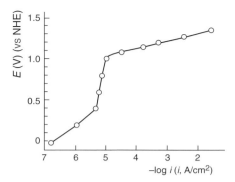

Figure 5.4 Anodic polarization curve on stainless steel 18-10 in a solution for producing sodium perborate (150 g/l Na_2CO_3, 25 g/l $NaHCO_3$ and 7 g/l B_2O_3). Temperature 15°C.

rate of the stainless steel can be attributed to the same causes that were discussed when the corrosion behavior of carbon and stainless steels in alkaline solutions were considered at the potentials of oxygen evolution.

Thus, under the considered conditions, stainless steel 18-10 maintains its passive state up to a potential of 1.8 V. Thereafter, electrochemical process of the oxidation of hydroxyl ions takes place on the metal surface at a fairly high velocity, and practically all the current is spent on this process. These results explain how piping and equipment made of stainless steel maintain their corrosion stability in the zone of action of leakage currents in the electrochemical perborate production plant.

5.2.3.   Acid sulfate containing electrolyte for copper electrorefining

Stainless steel 18-10 maintains its corrosion stability in dilute solutions of sulfuric acid up to a concentration of 10% at room temperature, if there is no influence of any other additional corrosion factors [22]. As the temperature increases, the field of its stability further decreases.

In the electrolyte for copper electrorefining, at a sulfuric acid concentration of 130 g/l and a temperature of 65°C, steel 18-10 is passive, owing to the presence of divalent copper, which plays the part of an oxidizer [22]. Corrosion tests have shown that in a solution of pure sulfuric acid, at the same concentration and temperature, the open circuit potential of steel 18-10 is equal to −0.095 V and the corrosion rate exceeds 30 g/m² h. It is seen from Figure 5.5 (curve 1) that steel 18-10 actively corrodes at this potential value and its passivation potential is above 0.1 V.

In a solution containing 130 g/l $CuSO_4$, which is similar to the technological solution of copper electrorefining, the open circuit potential of stainless steel is much more positive (0.6 V). At this potential value, stainless steel is in a passive state, which is proved by the results of the corrosion tests. After 5 h long tests, its corrosion rate did not exceed 0.04 g/m² h. Thus, the presence of Cu^{2+} ions in the solution is the cause of the steel passivation.

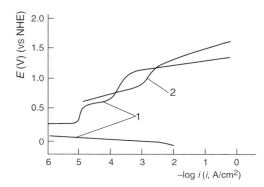

Figure 5.5 Anodic polarization plots on stainless steel 18-10 in solutions: 1 – 130 g/l H₂SO₄; 2 – 130 g/l H₂SO₄ + 130 g/l CuSO₄. Temperature 65° C.

The bends that are seen on curves 1 and 2 (Figure 5.5) at potentials 1.1–1.2 V are related to stainless steel transpassivation, which occurs as a result of the formation of compounds with the highest metal valence [23].

Corrosion tests at the galvanostatic polarization of steel 18-10 carried out in the technological solution of copper electrorefining have shown that at current densities of 0.2, 1 and 20 mA/cm², the metal potential shifts, respectively, to the values 1.2, 1.3 and 1.5 V and the corrosion rates are, respectively, 0.6, 4.6 and 117 g/m² h. A rough estimation indicates that the anodic current is mainly spent on the metal dissolution. Thus, admissible values of the metal potential under these conditions are in the range from the open circuit potential (0.6 V) to the transpassivation potential (1.1 V).

The transpassivation potential is reached at low current densities of less than 1 mA/cm². Therefore, the potential shift to the transpassivation area under attack by an external anodic current is the most probable.

Cases of corrosion damage were formerly noted in the areas of flange connections on the stainless steel 18-10 piping. Weld seams and crevices under the gaskets are located in these areas. For this reason, corrosion tests were carried out on specimens with weld seams and artificial crevices.

In galvanostatic (0.2 mA/cm²) corrosion tests of specimens with artificial crevices, no difference was revealed between the corrosion behavior of the metal inside a crevice and in the bulk of the solution.

In galvanostatic tests of specimens with weld seams at the anodic current densities 0.2, 1 and 20 mA/cm², the potential values that were installed on the specimens with and without weld seams were similar. However, after the tests, the weld seams and heat–affected zones were clearly seen on the welded specimens. These areas had undergone preferential etching. This was caused by structure–phase transformations that had taken place in the metal during the process of welding, which led to electrochemical irregularity of the weld seams and to a reduction in their corrosion stability with respect to the basic metal [24].

It can be concluded that attack by external anodic currents causes the corrosion of stainless steel 18-10 in sulfuric acid solutions as a result of transpassivation, and stimulates corrosion damage of the weld seams and of the thermally–affected zones.

REFERENCES

1. L. L. Shreir, R. A. Jarman and G. T. Burstein eds, Corrosion, Volumes 1 and 2, 3rd edn, Butterworth, 1994.
2. H. Kaesche, Die Korrosion der Metalle. Physikalisch-chemische Prinzipinen und aktuelle Probleme (in German), Springer-Verlag, Berlin – Heidelberg – New York, 1979, 388 S.
3. F. Todt, Korrosion und Korrosionsschutz (in German), Walter De Gruyter & Co., Berlin, 1961, p. 1427.
4. B. N. Kabanov, Electrochemistry of Metals and Adsorption (in Russian), Nauka, Moscow, 1966, 222p.
5. A. M. Sukhotin ed., Electrochemistry Handbook (in Russian), 1981, Khimiya, Leningrad, 486p.
6. L. M. Yakimenko, I. D. Modilevskaya and Z. A. Tkachek, Electrolysis of Water (in Russian), Khimiya, Moscow, 1970, 263p.
7. V. G. Beliaev, I. V. Parputz, V. I. Artemiev and A. M. Sukhotin, Zashchita Metallov, Vol. 20, No. 6, 1984, 914–916.
8. A. K. Agrawal, K. G. Sheth, K. Poteet and R. W. Staehle, J Electrochem. Soc., Vol. 119, 1637–1644.
9. M. N. Fokin, V. K. Juravlev, A. V. Mosolov et al., NaOH Influence on the Tendency of Stainless Steels to Local Corrosion in Alkali Media. In collected articles "Corrosion Protection in Chemical Industry", Vol. 5, NIITEKhim, Moscow, 1977, p. 3–8.
10. U. F. Frank and K. Weil, Zeitschrift fur Electrochem., B. 56, No.8, ss. 814–822.
11. V. M. Novakovsky, Review "Itogi Nauki i Tekhniki, Korroziya i Zaschita ot Korrozii", Vol. 2, VINITI, Moscow, 1975, pp. 5–26.
12. I. V. Riskin and A. V. Turkovskaya, Pitting Corrosion of Steels Kh18N10T and St10 under the Conditions of Solution Flow and Heat Exchange. Collected articles: "Corrosion of Chemical Equipment", Vol. 37, MIKhM, Moscow, 1971, pp. 3–13.
13. A. M. Sukhotin and A. L. Labutin eds, Corrosion and Protection of Chemical Equipment, Handbook (in Russian), Vol. 3, Khimiya, Leningrad, 1970, 308p.
14. L. V. Vaniukova and B. N. Kabanov, Dokladi ANSSSR, Vol. 59, No. 5, 1948, 917–920.
15. V. A. Andreeva, L. N. Shiganova and A. I. Krasilshchicov, Zashchita Metallov, Vol. 4, No. 3, 1968, 255–259.
16. A. M. Sukhotin and A. L. Labutin eds, Corrosion and Protection of Chemical Equipment, Handbook (in Russian), Vol. 6, Khimiya, Leningrad, 1972, 373p.
17. M. N. Fokin, A. V. Mosolov, V. K. Juravlev et al., Influence of Chlorate-Ions on Corrosion and Electrochemical Behavior of Stainless Steels in Hot Concentrated Alkali-Chloride Solutions.In collected articles: "Investigations of Metals Protection from Corrosion in Chemical Industry" Vol. 4, NIITEKhim, Moscow, 1976.
18. M. N. Fokin, V. K. Juravlev, A. V. Mosolov et al., Zashchita Metallov, Vol. 14, No. 6, 1978, 690–693.
19. Ya.M. Kolotirkin and G. M. Florianovich, Review "Itogi Nauki i Tekhniki, Korroziya i Zaschita ot Korrozii", VINITI, Moscow, 1975.
20. Handbook of Chemist (in Russian), 2 edition (revised), Vol. 3, Ed. by B. P. Nicolsky, Khimiya, Moscow – Leningrad, 1964, 443p.
21. L. M. Volochkova and A. I. Krasilshchicov, J. Fiz. Khimiyi, Vol. 23, No. 9, 441–444.
22. Ya.M. Kolotirkin and V. M. Kniajeva, J. Fiz. Khimiyi, Vol. 30, No. 9, 1956, 1990–2002.
23. N. D. Tomashov and G. P. Chernova, Passivity and Protection of Metals against Corrosion, Plenum Press, New York, 1967, 208p.
24. V. Cihal, Intercrystalline Corrosion of Stainless Steels (in Russian), Khimiya, Leningrad, 1969, 231p.

CHAPTER 6

CORROSION BEHAVIOR INVESTIGATIONS OF TITANIUM AND ITS ALLOYS IN THE MEDIA OF ELECTROCHEMICAL PLANTS, TAKING INTO ACCOUNT THE ATTACK BY ANODIC LEAKAGE CURRENTS

Contents

6.1. CORROSION AND ELECTROCHEMICAL CHARACTERISTICS OF TITANIUM

During the period of extensive growth of the application of titanium in aggressive media in chemical and other branches of industry, wide investigations of its corrosion and electrochemical behavior were carried out all over the world. Thanks to these investigations, a great amount of data has accumulated up to the present on the physicochemical and electrochemical properties of titanium which determine its corrosion resistance.

Titanium is a thermodynamically unstable metal: the ionization potential of the reaction $Ti = Ti^{3+} + 3e^-$ is equal to -1.23 V [1]. The corrosion stability of titanium is determined by its strong ability to passivate and by the high stability of the passive oxide film that is formed on its surface. Owing to the protective properties of this film, titanium possesses high corrosion resistance in many acid, neutral and alkaline media, including media containing activators such as chlor-ions [2].

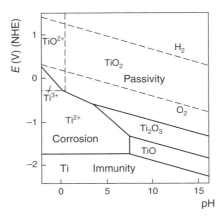

Figure 6.1 Simplified equilibrium potential–pH diagram for titanium in aqueous solutions at temperature 25°C [4].

The compositions and properties of the passive films that are formed on the surface of titanium depend on the pH, the composition and the temperature of the environment. The main oxides that a provide passive state to titanium are TiO, Ti_2O_3 and TiO_2 [3]. The areas of their stability are indicated in an equilibrium potential–pH diagram for titanium (Figure 6.1). In acid solutions, along with oxides, hydrides of various compositions, up to TiH_2, are formed under some conditions, which also exert an influence on the state of the metal surface.

In the presence of oxidizers, and under anodic polarization, a phase film of titanium dioxide TiO_2, which has a structure of rutile or anatase, is formed on the metal surface. As the potential and exposure time increase, the share of anatase in the oxide film increases [5].

Titanium belongs to the group of valve metals. The oxide film that is formed on its surface functions as a diode: it allows a cathodic current to pass through and almost completely cuts off the metal surface for the passage of an anodic current.

The presence of oxidizers, which is most typical for the aggressive media of electrochemical plants, does not reduce the corrosion stability of titanium, and, in most cases, even increases it. For example, in hydrochloric acid solutions, in the presence of dissolved chlorine, the potential of titanium shifts from the active state to the passive field, as in the case of anodic protection. As a result, the corrosion process on titanium stops [6]. Under similar conditions, i.e., in the simultaneous presence of strong oxidizers and activators in the solutions, structural metals, such as stainless steels corrode at extremely high rates.

Just by having the ability to maintain corrosion stability in media containing oxidizers and activators simultaneously, titanium found widespread application in electrochemical plants and, particularly, in chlor-alkali electrolysis plants, away from the zones of leakage currents attack. However, as shown in Chapter 4, in the action zones of leakage currents, titanium can undergo intensive corrosion.

Of particular interest is the corrosion and electrochemical behavior of titanium in the high value areas of anodic and cathodic potentials, to which the metal shifts as a result of attack by external currents of anodic and cathodic directions.

6.2. CHLORIDE AND CHLORIDE-ALKALI MEDIA OF CHLORINE-PRODUCING ELECTROCHEMICAL PLANTS

Most of the published results of investigations of the corrosion and electrochemical behavior of titanium were limited by the positive values of potentials 1.5–2 V, since they were carried out without considering the influence of external currents. In the absence of external currents, titanium potentials do not surpass these values. In chloride and sulfate solutions of all concentrations up to a temperature of 100°C and a pH of 1.5 (in the direction of its reduction), titanium has a stable passive state [2]. Some rise in the current density over the range of potentials 1.5–2 V was noted on the polarization curves obtained on a passive metal in NaCl solutions, which was explained in Ref. [7] as the result of chlorine evolution. In the long run, the polarization current density in the passive state goes down owing to the growth of an oxide film barrier.

The values of the activation potentials of titanium in NaCl solutions at temperatures of 20–100°C, in accordance with data cited by different authors [8–11], are within the range 15–5 V. Such a large range of potentials is explained by a profound dependence of the activation potential E_a, on the temperature. With the temperature increasing, the E_a shifts to the negative side [9–11]. The dependence of the E_a on the NaCl concentration and the pH is significantly lower. Under reduction of the pH from 7 to 1.5 and an increase in the NaCl concentration by 1–2 orders of magnitude, the E_a shifts to the negative side only by a few tenths of a volt [8, 11].

The values of the activation potentials of titanium are strongly dependent on the type of the ion activator. In KBr and KI solutions, the activation of titanium occurs at potentials which are significantly more negative than the activation potentials in chloride solutions: within the boundaries, 1.5–3 and 0.5–3 V, respectively [11]. The decisive influence of the anions on the E_a values indicates that the activation process occurs in accordance with the mechanism of pitting. It is interesting to note that the activating influence of the halogen ions increases in the order $I^- > Br^- > Cl^-$, while the strength of their influence on stainless steels is in the reverse order [12].

The decisive influence of the anions on the activation potential of titanium is proven by the fact that in chloride solutions the potential of the anodic activation of titanium is much more negative than in sulfate solutions, in which it attains 80–150 V [13, 14]. In phosphoric acid, the activation potential of titanium is above 400 V [15]. Destruction of the protective film at such high values of activation potential takes place as a result of its electrical breakdown [4].

Sulfate ions suppress the activating influence of chlor-ions, not only in the case of metals such as stainless steels [16], but also in the case of titanium. In accordance with the data in Ref. [17], as a result of a variation in the $NiSO_4/NiCl_2$ concentration ratio from 176/99 to 220/61.5, the activation potential of titanium in its electrochemical polishing solutions increased from 10.7 to 60 V. An increase in the activation potential of titanium when sulfate was added to chloride solution was also noted in Ref. [18]. These data show that in each specific case, special studies are necessary to assess the state of titanium in the presence of an external anodic current.

In the mid 1970s, work was performed by Mazza [19] which was assumed to be unique, in which the anodic activation of titanium by an external anodic current under electrolysis conditions was considered. The results of investigations on the corrosion behavior of titanium in hot (80°C) concentrated (310 g/l) NaCl solution, at high anodic potentials, are given in this work. These results were obtained under conditions of crevice formation under the deposits that formed on the metal surface. The formation of these crevices leads to a shift in the E_a from 10 to 7 V. Mazza explains the facilitation of the metal activation by acidification of the solution inside the crevices, which occurs as a result of the occurrance of the anodic process and hydrolysis of the anodic reaction products, as well as the rise in the chlor-ion concentration. It should be noted that also in later works [20], the elevated tendency of titanium to activation inside crevices under the conditions of anodic polarization was noticed.

Anodic potentiodynamic (2.88 V/h) polarization curves obtained on titanium in chloride (curves 1 and 2) and chloride-alkali (curve 3) solutions are shown in Figure 6.2. It can be seen, in accordance with above-cited data [7], that an increase in the current density, which is connected with the anodic process of chlorine evolution and simultaneous formation of an oxide film barrier, was observed within the range of potentials 1.6–2 V. At a potential of 2 V, the current density attained in the concentrated NaCl solution was about double that attained in the dilute solution, but further on, this difference slightly diminished. On increasing the potential further, a slow rise in the current density was observed up to the activation potential E_a, which is indicated by arrows. Somewhat lower values of the current density in the concentrated chloride solution at potentials above 2 V may be related to the significantly lower concentration of water molecules, which are donors of oxygen for the oxide film. Moreover, data given in Ref. [21] on the slowing down of the anodic processes on a passive metal under some conditions, when anions are adsorbed on the metal surface, must be taken into account.

In the considered field of potentials, the anodic current was mainly spent on the formation of the oxide film [22], which plays the part of a barrier layer. This was proved in experiments with a long-time (5 h) exposure of specimens to

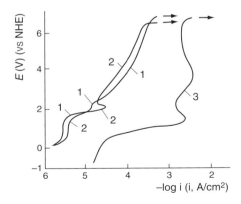

Figure 6.2 Anodic polarization plots on titanium in solutions of 1 and 300 g/l NaCl (plots 1 and 2, respectively) and in a solution of 120 g/l NaOH + 200 g/l NaCl (plot 3). Temperature, 90°C.

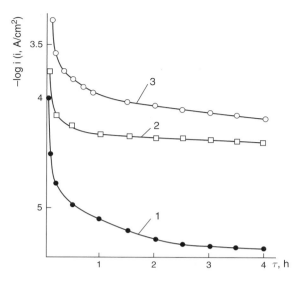

Figure 6.3 Chronograms of current density at potentiostatic polarization of titanium in a solution of 1 g/l NaCl at potentials (V): $1-2$; $2-4$; $3-6$. Temperature, 90°C.

constant potentials (Figure 6.3). During this exposure time, the current density almost stabilizes and at a potential of 6 V diminishes by 8–9 times, and at a potential of 2 V by more than one order of magnitude. The more positive the polarization potential, the faster the growth of the barrier film and, consequently, the faster the reduction in the current density. At the same time, the stabilized value of the current density at a potential of 6 V is an order of magnitude higher than at a potential of 2 V.

In chloride-alkali solutions, within the range of potentials of 0.5–1.5 V, a Taffel character is observed of the dependence of the current density on the potential (Figure 6.2, curve 3). The only possible anodic process that can take place in this area of potentials is oxygen evolution. The values of the anodic current densities at potentials from 0.5 V up to the activation potential are 1–2 orders of magnitude higher than those in chloride solutions. In Refs. [23, 24], it is suggested that this effect is related to different mechanisms of oxygen evolution in neutral and alkaline solutions. In the latter case, oxide barriers are not formed on the metal surface during the anodic oxidation of hydroxyl ions.

Values of the activation potential in solutions of 1 and 300 g/l of NaCl at a temperature of 90°C are equal, respectively, to 6.2–6.4 and 5.8–6.0 V. An increase in the chloride concentration by more than two orders of magnitude leads to a shift in the potential to the negative side by only 0.2–0.4 V.

In a chloride-alkali solution, E_a is equal to 6.5–6.6 V, i.e., it is quite close to the values of the activation potential in neutral solutions of chloride.

Limited studies were carried out to estimate the influence of the dissolved chlorine on the anodic behavior of titanium in a dilute (1 g/l) solution of NaCl. A condensate of a similar composition is formed inside the wet chlorine lines in chlor-alkali plants. The open circuit potential of titanium in a solution saturated with chlorine (by bubbling chlorine through the solution for 1 h) is 0.88 V, i.e., it is much more positive than in the absence of dissolved chlorine (−0.06 V). In a chlorine-saturated solution, the value of the potential E_a is about 0.2 V more negative than in a solution without chlorine. This may be explained by the acidification of the chlorine-saturated solution to pH 1.6–2 as a result of chlorine hydrolysis. Acidification with hydrochloric acid of a solution that does not contain chlorine to pH 1.6 also leads to a shift of the E_a to the negative side by about 0.2 V. These results show that dissolved chlorine does not influence the protective properties of the oxide film at the higher values of the activation potential. Moreover, they are in line with the data cited above on the low dependence of E_a on the concentration of chlor-ions and the pH [11].

Owing to the dynamic character of the polarization curves presented in Figure 6.2, the values of the current densities at which anodic activation of the metal is attained are exaggerated. To find more exact values of the current densities at which the E_a is attained in chloride and chloride-alkali solutions, chronograms of the potentials were studied at different current densities (Figure 6.4). The minimal current density at which E_a was attained in 1 g/l NaCl solution was equal to 0.2 mA/cm^2 (curve 1). As the current density increased, the rate of the shift in the potential to the positive side also increased: at 0.2 mA/cm^2, E_a was attained after 3 h of polarization, and at 1 mA/cm^2, after as short a time period of 5–10 min (curves 1–3). The E_a values obtained in the galvanostatic regime were slightly more positive than the values obtained in the potentiodynamic regime of polarization (see Figure 6.2).

In 300 g/l NaCl solution, at a current density of 0.1 mA/cm^2, E_a was reached after 30–40 h of polarization. This shows that titanium activation in neutral and mildly acidic chloride solutions occurs at low values of anodic current density.

At the same time, in a solution of 120 g/l NaOH + 200 g/l NaCl, at a current density which is an order of magnitude higher (1 mA/cm^2), the

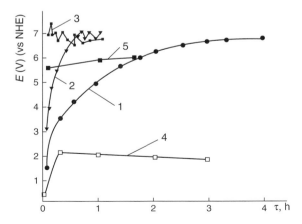

Figure 6.4 Chronograms of titanium potential in solutions of 1 g/l NaCl (plots 1–3) and in a solution of 120 g/l NaOH + 200 g/l NaCl at current densities (mA/cm^2): 1 − 0.2; 2 − 0.5; 3, 4 − 1; 5 − 10. Temperature, 90°C.

potential is stabilized at a value of 2.4 V (curve 4) and the titanium does not undergo corrosion. This confirms once more the above-mentioned suggestion as to the higher conductivity of the oxide film that is formed in alkaline solutions. However, at a current density of $10 \, mA/cm^2$, the titanium potential shifts instantly to the area of activation (curve 5).

A drastic proliferation of current takes place when the activation potential E_a is reached under potentiodynamic polarization (see Figure 6.2). As anodic activation is attained in the galvanostatic regime of polarization, oscillations in the potential are observed around the potential E_a (Figure 6.4). In all cases, after the metal exposure at the potential E_a, pits filled with white corrosion products of titanium dioxide are formed on the metal surface.

It was already noted that anodic activation of titanium in the presence of chlor-ions proceeds according to the mechanism of pitting. Hence, under the given conditions, the activation potential E_a of titanium has to be considered as a critical pitting potential. Under the potentiostatic regime of polarization, metal activation does not occur below the E_a value, and only current reduction is observed (see Figure 6.3). Even metal exposure to a potential of 6 V, only 0.2 V below the E_a value, does not lead to metal activation.

Assessment of the weight loss of titanium specimens after polarization at constant current densities, under the conditions of pitting formation, has shown that about 70% of the current is spent on metal dissolution (taking into account the Ti^{4+} ions). The rest of the current is spent on the anodic evolution of gases (chlorine and oxygen) and on titanium oxidation at high anodic potentials [18].

6.3. ELECTROLYTE FOR COPPER ELECTROREFINING

Electrolytes for copper electrorefining belong to the most typically aggressive environments of electrochemical plants, producing pure metals by their deposition from aqueous solutions. Their aggressiveness to structural metals is explained by the high acidity and elevated temperatures (60–70°C). The compositions of the elec-trolytes used for copper electrorefining do not differ significantly in different countries [1]. The results of studies carried out with a solution of average composi-tion conforming to the electrolyte of a typical copper electrorefining plant (see Table 3.3), are considered below.

There are data [2] indicating that titanium possesses corrosion stability in dilute sulfuric acid at concentrations of up to 5% at room temperature and only up to 0.2% at 100°C. Corrosion studies have shown that the corrosion rate of titanium in a pure solution of 130 g/l H_2SO_4, at a temperature of 60°C, is very high, 37 g/m^2 h, and the open circuit potential of the corroding titanium is equal to -0.465 V. At the same time, in the electrorefining electrolyte, the open circuit potential is stabilized at a value of 0.6 V, and the weight losses are close to zero.

An area of active metal dissolution is seen in the anodic polarization curve obtained on titanium in a solution of 130 g/l H_2SO_4 at a temperature of 60°C

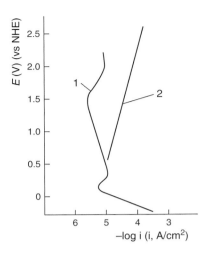

Figure 6.5 Anodic polarization plots on titanium in solutions: $1 - 130\,g/l$ H_2SO_4; $2 -$ technological solution of copper electrorefining. Temperature, $60°C$.

(Figure 6.5, curve 1). The potential of complete titanium passivation in this solution is equal to 0.1 V, i.e., it is about 0.5 V more negative than the open circuit potential in the technological solution. As in the case of stainless steel, titanium passivation is caused by Cu^{2+} ions, which raise the oxidation–reduction potential of the technological solution.

Upon a further increase of the potential, a slow growth of current takes place (curve 2), which is mainly spent, as noted above, on oxygen evolution and oxide film formation [13].

Cylindrical specimens, 1, of diameter 10 mm and working surface area $8\,cm^2$ were tested under conditions of galvanostatic polarization. Specimens of this kind are shown in Figure 6.6a. Their connection with the power source was through a metallic rod, 2, isolated from the solution by a glass tube, 3, and a sealing gasket of Teflon, 4. In the considered works, the specimens were studied without the rubber ring shown in Figure 6.6a. At a current density of $20\,mA/cm^2$, the potential of the titanium shifted to 150–155 V (Figure 6.7, curve 1) and a thick gray film which increased the mass of the specimen by $1\,mg/cm^2$ formed on the metal surface. In a similar, repeat 5 h long polarization test on the previously oxidized titanium specimen, a similar character of dependence of the potential on time was observed. However, damage of a local character occured at the edges and under the Teflon gasket. The total surface area of the damage was $\sim 0.02\,cm^2$, and the weight losses were about $1\,mg/cm^2$.

At a current density of $50\,mA/cm^2$, the potential reached values of 140–150 V over 5 h of polarization (curve 2), and after that shifted slightly to the negative side and varied in the range 130–140 V. In this case, the damaged surface area and the weight losses increased, respectively, to $0.06\,cm^2$ and $\sim 8\,mg/cm^2$. The damage

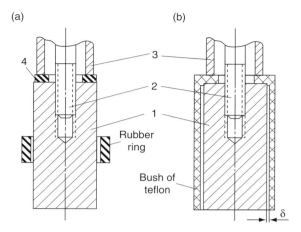

Figure 6.6 Specimens with artificial crevices produced by: a – rubber ring; b – bush of Teflon. 1 – specimen; 2 – metallic rod (current lead); 3 – isolating glass tube; 4 – Teflon gasket.

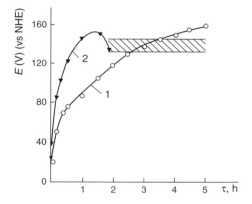

Figure 6.7 Chronograms of titanium potentials in technological solution of copper electrorefining under galvanostatic polarization by anodic current densities (mA/cm^2): 1 – 20; 2 – 50. Temperature, 60°C.

tended to localize close to the Teflon gasket, but also appeared on other areas on the surface of the specimens.

High values of the activation potential E_a indicate that under the conditions of interest, the activation occurred as a result of electrical breakdown of the oxide film on the titanium. Particular damage at the edge, close to the gasket, was due, most probably, to the effect of film spalling at the edge of the area and, in addition, to the stress produced by pressing the gasket onto the specimen. However, these effects are slight, which is proved by the fact that in both cases the values of E_a were close to each other. In galvanostatic studies of spade–shaped specimens bent at 90° (to produce internal stresses), the E_a value was also not significantly changed.

6.4. MEDIA OF NICKEL ELECTROREFINING

Sulfates and chlorides, the components that form a part of the composition of the electrolytes (catholyte and anolyte) of nickel electrorefining, are typical for aggressive media of the chemical and non-ferrous metallurgical industries. The aggressiveness of the catholyte and anolyte is mainly determined by a high concentration of chloride, a low pH and an elevated temperature of these solutions.

Studies were carried out in a solution conforming to a typical catholyte composition in a nickel electrorefining plant. Some experiments were done in an anolyte – a solution which included, along with the components presented in Table 3.3, ions of Cu^{2+} (1 g/l) and Fe^{3+} (0.8 g/l).

Corrosion tests lasting 30 h showed that in the absence of an external current, titanium maintains its corrosion stability in the considered solutions. During the time of testing, the specimens did not lose their metallic shine. The open circuit potential, which stabilized after 1 h of immersion of the specimens into the solution, was equal to 0.35 V in the catholyte and 0.7 V in the anolyte. The higher potential value in the anolyte is explained by the presence of Cu^{2+} and Fe^{3+} ions in this solution.

At the potentiodynamic polarization (0.72 V/h), the beginning of a rise in the current, which was related to the evolution of oxygen and chlorine on surface of the titanium, was noted at potentials 1.3–1.4 V (Figure 6.8). Above these potentials, the current density was significantly higher than those in pure chloride or sulfate solutions. For example, at a potential of 3.8 V, a current density of about 50 mA/cm^2 was reached, and a dark-yellow oxide film formed on the metal surface.

The results of potentiostatic oxidation at potentials 2, 3 and 4 V are displayed in Figure 6.9. Initially (0.5 h), the current density went down to values that were 2–3 orders lower than that at potentiodynamic polarization. However, after this period of time, a new current density growth occurred, which indicated some changes in the structure of the oxide film during the process of a long-time the polarization. As the polarization potential was increased from 2 to 4 V, the current density in the passive state increased by one order of magnitude.

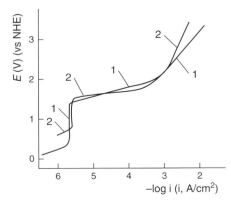

Figure 6.8 Anodic polarization plots on titanium in solutions of nickel electrorefining: 1 – catholyte; 2 – anolyte. Temperature, 90°C.

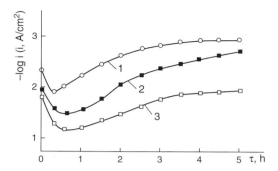

Figure 6.9 Chronograms of current density on titanium in catholyte of nickel electrorefining at potentials (V): 1 – 2; 2 – 3; 3 – 4. Temperature, 85°C.

Under the conditions of long-time galvanostatic polarization, the titanium potential shifted to the positive side, up to the value of the activation potential E_a equal to 80–85 V (Figure 6.10). Destruction of the passive film, which occurs at this potential, is accompanied by intense local corrosion of the titanium. The time taken to reach the value of E_a depends on the density of the polarization current. When it was increased from 5 to 50 mA/cm^2, the time went down from 25–30 to 10 h. Two areas are notable on the obtained curves. Area A (0.5–1 h) is characterized by a fast shift to some potential value. The higher the current density, the higher this potential. At the area B, the potential slowly shifts up to the value of E_a.

The obtained values of E_a are in compliance with the data given in work [17], in which the results of studies of titanium in chloride–sulfate solutions are presented. The more negative values of E_a in these solutions, with respect to sulfate solutions of copper electrorefining, are explained by the presence of chlor-ions and by the higher solution temperature. The influence of the chlor-ions indicates that these ions

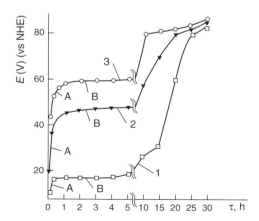

Figure 6.10 Chronograms of titanium potential in catholyte of nickel electrorefining at current densities (mA/cm^2): 1 – 5; 2 – 20; 3 – 50. Temperature, 85°C.

are involved in the act of metal activation. This suggests that anodic activation of titanium in a chloride–sulfate solution cannot be considered as a result of electrical breakdown, despite the relatively high E_a value.

Moreover, the results of experiments under conditions of galvanostatic polarization point to a high susceptibility of the activation potential E_a to the crevice effect and to the factors connected with the state of the metal surface. In some experiments, metal activation occurred at potentials that were significantly below 80 V. Upon inspection of these specimens, after testing, it was revealed that the area of the edges (where the defects in the passive film are most probable) and the area close to the Luggin capillary (where elevated chloride concentration and crevice effects are possible) are most often damaged. In this connection, investigations of the susceptibility of titanium to the presence of welding seams and crevices were carried out. The most accessible and close criterion to the actual conditions for the estimation of this influence is the change in the activation potential E_a in the presence of crevices and welding seams.

6.5. Influence of Welding Seams and Crevices

The presence of welding seams and crevices under gaskets is most typical for flange connections of piping and nozzles of different kinds of equipment. As already noted, these areas usually coincide with terminal parts of tubes and other structures and are therefore most often attacked by leakage currents. As the corrosion resistance of the metal can be the lowest in the areas of the welding seams and crevices, the influence of these structural-technological factors must first of all be taken into account in the presence of leakage currents.

6.5.1. Welding seams

The corrosion resistance of titanium welding seams is in most cases, close to the corrosion resistance of the basic metal [2]. However, there are data [25, 26] to prove that in environments containing strong oxidizers (nitric acid, chlorine dioxide, mixtures of chromic and hydrofluoric acids, etc.) severe corrosion of titanium welding seams and of heat-affected areas develops. It is suggested in Ref. [26] that the cause of corrosion damage of welding seams is the preferential dissolution of titanium. The β-phase is formed in the zone of a welding seam as a result of the heating–cooling cycles, and it contains elevated concentrations of iron. The corrosion stability of the β-phase is, in general, higher than the stability of the α-phase [27]. The reduction of the passivating properties of the β-phase under the considered conditions is connected to its elevated iron content.

A change in the welding seam structure, leading to the formation of a metastable, coarse-needled α-phase, similar to martensite, may also be a cause of the reduction in the corrosion stability of the welding seams. The lattice parameters of this phase slightly differ from the parameters of the α-titanium lattice [27]. Moreover, saturation by oxygen, nitrogen and hydrogen in the process of welding also reduces the corrosion stability of the titanium welding seams [27]. Hydrogen has the most negative influence because it promotes porosity of the welding seams [28].

The destruction of the welding seams can be promoted by uneven heating of the metal during the welding process which produces internal stresses in the superficial metal layer and hinders the formation of a perfect and regular passive film.

As the influence of the external anodic current is, in some respects, similar to the influence of oxidizers, it was necessary to study its influence on the corrosion stability of welding seams. Titanium test specimens in the form of 1 mm thick plates were produced by welding two similar pieces together to give a total working surface area of $6\,cm^2$. The welding beads formed after the argon–arc welding were ground, to obtain flat specimens. Other operations of the surface treatment of the specimens before testing (degreasing, washing and drying) were carried out as usual.

The value of the activation potential E_a of the welded titanium in 1 g/l NaCl solution at temperature of 90°C decreased by only $\sim$0.2 V (to 6–6.2 V), but the damage was concentrated preferentially in the areas of welding seams and heat-affected zones. These results indicate that, in general, the corrosion stability of a welded titanium construction in chloride solutions is insignificantly reduced under attack by external anodic currents, but that it is necessary to avoid the presence of the welding seams in areas where the leakage currents may have maximal values.

Under galvanostatic polarization of the welded titanium specimens in the solution of copper electrorefining at a current density of $5\,mA/cm^2$, the potential E_a dropped drastically, from 140–150 to 9.5–10 V (Figure 6.11, curve 1).

A drastic reduction of the E_a, from 80–85 to 5.4–5.6 V, under galvanostatic polarization of welded titanium specimens in the nickel electrorefining catholyte also took place (Figure 6.11, curve 2).

The potential E_a was attained in both cases during the first half hour of polarization. The damages was clearly limited to the welding and heat-affected zones and was not found outside these zones.

The possible causes of the reduction in the corrosion stability of titanium welding seams were considered above. As a result of a reduction in the metal passivation ability on the welding seam and in the heat-affected zones, the major part of the external current is concentrated in the areas of the damaged passive film, and the local current density in these areas becomes higher than those on the rest of the

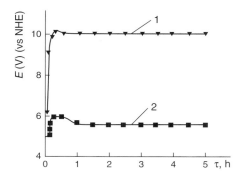

Figure 6.11 Chronograms of potential of welded titanium specimens in electrorefining solutions of: 1 – copper; 2 – nickel. Temperatures, respectively, 60 and 85°C.

metal surface. This might be the cause of the high susceptibility of the welding seams to attack by external anodic currents.

It should be noted that the E_a values of welded titanium in chloride and in chloride–sulfate solutions are close to each other. As noted above, sulfate ions suppress the activation effect of chlorine ions [17, 18]. It may be presumed that in the area of a welding seam, where the local value of the current density is elevated, such a change in the ratio of the concentrations of the Cl^- and SO_4^{2-} ions occurs where the SO_4^{2-} ions do not suppress the activation effect of the Cl^- ions. Data [17] on reducing the E_a from 60 to 10.7 V by changing this ratio, point to such a possibility. These E_a values are of the same order as the E_a values obtained on titanium and on titanium welding seams in the nickel electrorefining solution. As was already suggested, the low E_a value may indicate that the most possible mechanism of titanium activation is pitting. Thus, in the area of the damaged passive film and, consequently, of elevated local current density, the concentration of chlor-ions increases and this leads to a breakdown of the passive film at a much more negative potential.

6.5.2. Artificial crevices

Crevice corrosion of titanium, in the absence of attack by internal currents, is well-studied. It has been found that the protective properties of the passive film assure the corrosion stability of titanium structures in the areas of crevices and gaps in most neutral environments, including chloride-containing media, at temperatures of up to 100°C [2]. Above100°C, titanium is prone to crevice corrosion in neutral solutions; in acid solutions crevice corrosion occurs at lower temperatures [29–33]. Depletion of the solution inside the crevices by oxygen is one of the major causes of titanium activation. Therewith, the crevice effect occurs as a result of acidity elevation [29–31] or due to the accumulation of Ti^{3+} ions, which reduce the redox potential of the solution. Japanese researchers explain the crevice corrosion of titanium in concentrated chloride solutions by salt deposition inside the crevice and by depletion in the solution of water molecules, which play the part of a passivation agent [33]. As cited above (see Chapter 5), American researchers also explain the crevice corrosion of titanium in wet chlorine and chlorinated water by dehydration of the solution.

Investigations of crevice corrosion under attack by external anodic currents were carried out under galvanostatic polarization of 20 mm long cylindrical specimens of 10 mm diameter. The crevice effect was produced with the help of a 4 mm wide rubber ring of internal diameter 8 mm. The ring was placed on the middle part of the specimen (Figure 6.6a). Since the tension of the ring was low, a crevice was produced between the ring and the specimen. The ratio of the surface area inside the crevice to the open surface area of the specimen was 1:5. Experiments were carried out in sulfate and in chloride–sulfate solutions (electrolytes of copper and nickel electrorefining).

The estimation of the influence of the gap width in a chloride–sulfate solution was also carried out on cylindrical specimens that were mounted inside a bush made of Teflon (Figure 6.6b). The gap width δ between the specimen and the internal surface of the bush was regulated in the range 0.1–2 mm by mounting bushes of different internal diameters. The ratio of the surface areas inside the crevice and on the open surface (end face) of the specimen was in this case equal to 6:1. Sealing between the specimen and the glass tube isolating the metallic rod (current lead) from the solution was provided by a collar disposed at the upper end of the Teflon bush (see Figure 6.6b). A similar method of testing crevice corrosion is described in Ref. [30].

Studies were carried out at current densities of 5–50 mA/cm^2 in acid sulfate and of 1–5 mA/cm^2 in chloride–sulfate solutions. The duration of the experiments on specimens with and without crevices was, respectively, 5 h and 5–30 h.

Experiments in 1 g/l NaCl solution were executed only on cylindrical specimens without artificial crevices. The development of the crevice effect was in this case taken into account when it was fixed under the sealing gasket, 4, between the specimen, 1, and the glass tube, 3 (see Figure 6.6a).

The same experiments were carried out in a solution that had the same composition as the nickel electrorefining catholyte, but without chlor-ions. Specimens without artificial crevices were also tested in a chloride–sulfate solution, in which the NiCl$_2$/NiSO$_4$ ratio was equal to 120/171.

As already noted, in chloride solutions at temperatures below 100°C, including chlorine saturated solutions, no corrosion of titanium was observed in the absence of external anodic current. When the potential E_a was attained under a long-time galvanostatic polarization at an anodic current density of 0.2–1 mA/cm^2, the major damage was concentrated around the gasket which was pressed to the upper end surface of the specimen (Figure 6.12). This indicates the crevice effect of titanium in chloride solutions under attack by an anodic current. On the other hand, as seen from Figure 6.12, the damage appeared not only

Figure 6.12 Corrosion in the crevice under the gasket and outside the crevice, at the end face of a cylindrical titanium specimen, after anodic polarization in 1 g/l NaCl solution. Temperature, 90°C. Magnification × 8.

around the crevice, but also on other areas of the metal surface. Hence, it may be assumed that under the conditions of interest, the difference in the E_a values inside the crevice and on the open surface of the metal is insignificant. The observed crevice effect is, most probably, stimulated by an increase in the chlor-ions and reduction of the pH inside the crevice. As shown, the pH value and chloride concentration exert a slight influence on the E_a value of titanium in a chloride solution, which explains the low exertion of the crevice effect on the titanium in this solution.

The open circuit potential of titanium specimens in a copper electrorefining solution, with or without crevices, was equal to 0.6 V. In 30 h long corrosion tests, the specimens with crevices produced by rubber rings did not undergo crevice corrosion.

In similar tests with the impression of an external anodic current, it was found that the crevice effect is low. At a current density of 20 mA/cm^2 the potential attained 110 V in 5 h and continued to shift to the positive side. A gray oxide film formed over the entire specimen surface, except for the surface located under the rubber ring, and the weight of the specimen increased by ∼1 mg. The surface under the ring retained its metallic shine. This may be explained by the hindrance to the growth of the oxide under the conditions of limited transfer of oxygen from the bulk of the solution into the crevice under the ring. Nevertheless, under the ring, no corrosion damage was noticed. At a current density of 50 mA/cm^2, the potential shifted to the values 130–135 V and corrosion damage occurred only under the rubber ring (Figure 6.13, specimen 1) and did not propagate to other areas of the metal surface. The E_a values under these conditions were about 5 V more negative inside the crevice than on the open surface of the metal. Against the background of such a high E_a value, its reduction by 3–4% can be considered insignificant. However, this effect must be taken into account in the copper electrorefining electrolyte, where the flange connections and other areas of the titanium structures, where crevices can be formed, are located only in the zones of attack by the maximal leakage currents.

The potential–time dependence under the galvanostatic polarization of titanium specimens with crevices in the copper electrorefining solution is presented in Figure 6.14 (curve 1). It is similar to the dependence obtained in the absence of

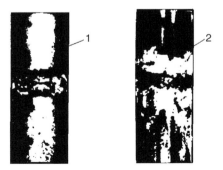

Figure 6.13 Corrosion damage of titanium specimens inside crevices produced by rubber rings after anodic polarization in solutions of: 1 – copper electrorefining, current density 50 mA/cm^2; 2 – nickel electrorefining, current density 5 mA/cm^2. Temperatures, respectively, 60 and 85°C.

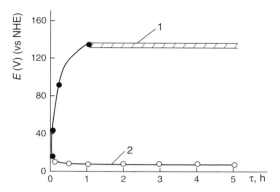

Figure 6.14 Chronograms of potential of titanium specimens with artificial crevice, produced by a rubber ring, under polarization in solutions of electrorefining: 1 – copper; 2 – nickel. Polarization current densities, respectively, 50 and 5 mA/cm². Temperatures, respectively, 60 and 85°C.

crevices. In both cases, activation occurs as a result of electrical breakdown of the protective oxide film.

In corrosion tests on titanium specimens with artificial crevices in the nickel electrorefining catholyte, no crevice effect was observed. The open circuit potential of the specimens with and without crevices was equal to 0.35 V. Under galvanostatic polarization of specimens with a rubber ring at a current density of 5 mA/cm², the potential instantly shifted to 8 V and then stabilized in about 15 min at values of 5.35–5.45 V. After 5 h long polarization, deep local damage occurred in the area of the crevice and on the open surface close to this area (Figure 6.13, specimen 2). The rate of local corrosion was above 240 mm/year. Thus, in the nickel electrorefining solution, titanium is very prone to crevice corrosion under attack by an external anodic current. In the presence of crevices, the potential E_a drops from 80 to 5.4 V.

On the specimens with an artificial crevice produced by a Teflon bush, at a gap width δ equal to 0.1–2 mm, titanium activation also occurred in a nickel electrorefining solution at low E_a values (from 5.2 to 9 V). The corrosion damage occurred at the mouth of the crevice and propagated to a distance of 5–7 mm inside the crevice. The concentration of the damage near the crevice mouth may be explained by the irregular potential and current distribution along the depth of the crevice. It is shown in Ref. [30] that the highest values of the potential and the current are reached at the crevice mouth and on the end face of the specimen and they decrease through the depth of the crevice.

Along with the surface inside the crevice, the end face of the specimens located on the open surface of the specimen was also damaged. The wider the gap δ (in the mentioned limits), the clearer the susceptibility to corrosion damage was the end face. At δ values of 0.1–0.2 mm, no corrosion of the end face was noticed; at $\delta = 0.5$ mm, damage was visible in the crevice mouth and on the end face; at δ values of 1 and 2 mm, the damage was preferentially concentrated on the end face. These results indicate that the crevice corrosion in the chloride–sulfate solution under attack by an external anodic current is caused not only by interference towards the formation of the oxide film but also due to a change in the composition of the solution inside the crevice. It can be assumed that metal activation begins inside the crevice. Aggressive components that accumulate inside the crevice reach adjacent areas on the open surface of the specimen and promote their activation. The wider the crevice, the greater the volume of the accumulated aggressive components and the easier they leak out of the crevice toward the adjacent areas of the open metal surface. Similar contamination by aggressive corrosion products leaking out of pits, which can also be considered as crevices [16], was observed under the conditions of the anodic polarization of stainless steel in NaCl solutions [34].

The elevated aggressiveness of the solution inside the crevice is, most probably, connected with the accumulation of chlor-ions and with an increase in the acidity of the solution [30]. This was proved by the fact that in a solution of a similar composition, but which did not contain chlor-ions, no crevice corrosion of titanium under anodic polarization by a current density of 5 mA/cm^2 was observed.

As in the experiments with welding seams, the E_a values obtained on the specimens with crevices in a chloride–sulfate solution are close to the E_a values found in solutions of pure chloride. It may be possible that under the conditions of anodic polarization, the ratio of the concentrations of chloride to sulfate inside the crevice increases. This, in turn, leads to a reduction of the effect of the sulfate in suppressing the activating influence of the chlor-ions. This cause of the E_a reduction inside the crevice, supported by the data of work [17], was already taken into account when the results of the testing of welded seams were considered. Results from studies of titanium specimens without crevices in a solution with an elevated concentration of NaCl (120 g/l), with all other things being equal, are also in line with this suggestion. The E_a in this solution was equal to 5.6–5.7 V, i.e., it was quite close to the E_a values obtained inside the crevice. It must also be taken into account that the activating influence of chlor-ions can increase when the acidity of the solution rises.

The drastic reduction of the activation potential E_a inside a crevice in a chloride–sulfate solution, under attack by an anodic current, resulting in the accumulation of chlorine ions and acidification of the solution inside the crevice passes, most probably, in accordance with the mechanism of pitting corrosion. At the same time, titanium activation at the open surface occurs, preferentially, as a result of an electrical breakdown of the passive film.

It is obvious that the presence of the crevices must be taken into account when the corrosion stability of titanium structures working under the considered conditions is estimated and when methods for the protection of these structures against electrocorrosion are developed.

In studies of spade-shaped specimens of titanium in the electrolytes of copper and nickel electrorefining, the elevated susceptibility of titanium to waterline corrosion was revealed under the conditions of anodic polarization. In the copper electrorefining solution, despite localization of the damage on the legs of the specimens, at the area of waterline the reduction of the E_a value was insignificant: the E_a was equal to 140 V. On the other hand, in the catholyte of nickel electrorefining, the waterline effect resulted in a drastic drop of the E_a, to values of 6.5–7 V. Thus, in nickel electrorefining solutions, the waterline, as well as the crevice, is a "weakened" area of the titanium structure under the conditions of attack by external anodic currents.

6.6. TITANIUM ALLOYS

Information on the corrosion stability of titanium alloys under attack by external currents is limited. In work [7], it is conveyed that alloy of titanium with aluminum and with a high content of vanadium (30–45%) were damaged under anodic polarization as a result of the preferential dissolution of aluminum and vanadium. On alloys containing 7.5 and 15% of vanadium, and on titanium alloys with wolfram, a current density equal to $0.4\,mA/cm^2$ was reached in the area of chlorine evolution, but no increase in the weight loss or potential with time was noticed. At higher current densities the potential shifted to values above 2 V.

As shown above (see Chapter 4), under the attack of leakage currents of an anodic direction, an alloy of titanium with 0.2% palladium was damaged more slowly than unalloyed titanium, since part of the external current was spent on chlorine evolution on the metal surface enriched with palladium.

Most interesting are the results of investigations of titanium–nickel alloys. It was shown in Refs. [22, 35] that in chloride solutions a significant rise (by about one order of magnitude) in the anodic current density was observed on these alloys in the area of chlorine evolution compared to unalloyed titanium. Studies of the anodic behavior of titanium–nickel alloys in the area above 2 V were not carried out.

According to these works, the increase in the anodic current density on titanium–nickel alloys is related to the formation of an intermetallic compound, Ti_2Ni, which increases the conductivity of the oxide film. The presence of this intermetallic compound hinders the formation of an oxide film barrier on the metal surface [22].

It was found [36, 37] that the barrier properties of the oxide film formed on the surface of the titanium–nickel alloy during anodic polarization depend on the thickness of the external oxide layer, which is composed of TiO_2 and does not contain nickel compounds. The higher the nickel content in the alloy, the thinner is this layer. The concentration of Ti_2Ni in the alloys increases from the internal boundary of the superficial layer of TiO_2 toward the depth of the metal.

A comparative estimation of anodic polarization curves obtained in the potentiodynamic regime (2.88 V/h) on different titanium alloys in 1 g/l NaCl solution was carried out to obtain preliminary information about the influence of alloying components on the values of the potentials E_a and the current densities in a passive state. Titanium alloys with 3 and 6% aluminum (AT-3 and AT-6), with 0.2% palladium (alloy 4200), with 5% tantalum (alloy 4204) and with 2% nickel were investigated (Figure 6.15).

Addition of 3 and 6% aluminum to titanium leads to a sharp shift in the E_a to the negative side, from 6 V for unalloyed titanium to values, respectively, of 4.9 and 2.6 V (curves 2 and 3). The current density on these alloys in the passive state is noticeably higher than that on titanium. In the area of oxygen and chlorine evolution, a significant rise in the current takes place that, in accordance with the data of Ref. [7], may be explained by selective dissolution of the aluminum and by oxygen and chlorine evolution on the activated metal surface. As the aluminum content in the superficial layer of the metal is reduced, the current density diminishes.

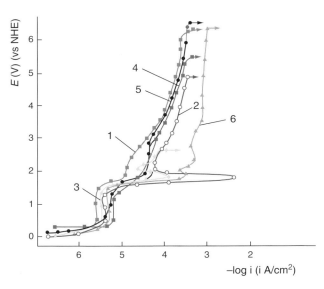

Figure 6.15 Anodic polarization plots in 1 g/l NaCl on titanium (1) and on its alloys with: 2 – 3% Al; 3 – 6% Al; 4 – 5% Ta; 5 – 0.2% Pd; 6 – 2% Ni. Temperature, 90°C (see color plate 1).

The E_a value of Ti–0.2% Pd alloy (5.5 V) is not that far from the E_a value of unalloyed titanium. This small difference in the E_a values can be connected not only to the presence of palladium, but also to the difference in the surface state of the titanium and of the alloy or with some other uncontrolled factors. The polarization curves on titanium and Ti–0.2% Pd alloy almost coincide, because the content of Pd in the passive oxide film does not noticeably influence the anodic process [38].

The E_a values on titanium alloys with 5% Ta and with 2% Ni are close to the E_a values on titanium. Current densities in the passive state obtained on titanium and on its alloy with 5% Ta almost coincide. In contrast, the current density on Ti–2% Ni alloy in the field of potentials above 1.5 V and up to the activation potential E_a is significantly higher (in some areas by one order of magnitude) than that on unalloyed titanium. According to Refs. [22, 35], which were already cited above, the rise of the current density on the passive alloy is attributed to the presence of the intermetallic compound Ti$_2$Ni in its superficial layer.

These results stimulated more detailed investigations of the corrosion and electrochemical behavior of Ti–2% Ni alloy, as the most promising for use under conditions of attack by external currents (Figure 6.16). At potentials above 1.5 V, the current density in 300 g/l NaCl solution is slightly lower than that in diluted (1 g/l) NaCl solution (curves 1 and 2).

Reduction of the current density on unalloyed titanium in concentrated NaCl solution, as compared with diluted NaCl solution, was already noted (see Figure 6.2). It was explained by the reduction in the concentration of water molecules, donors of oxygen for the oxide film, in the concentrated NaCl solutions. On the Ti–2% Ni alloy this effect is more significant, most probably, due to the reduction in the oxygen

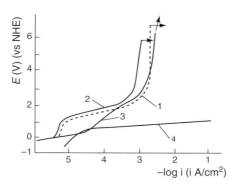

Figure 6.16   Anodic polarization plots on alloy Ti–2% Ni in solutions: 1 – 1 g/l NaCl; 2 – 300 g/l NaCl; 3 – 1 g/l Na$_2$SO$_4$; 4 – 120 g/l NaOH + 200 g/l NaCl. Temperature, 90°C.

overvoltage on the nickel-containing alloy. In conformity with the data in Ref. [39], this effect proves that at high potential values, preferential evolution of oxygen, but not of chlorine, takes place. This suggestion is supported by the fact that in solutions of 1 g/l NaCl and 1 g/l Na$_2$SO$_4$ (where the current is mainly spent on oxygen evolution) at potentials of 1.5–2 V, the values of the current densities are close to each other.

In 1 g/l NaCl solution, the activation potential E_a was reached on titanium after 3–3.5 h of polarization at a current density of 0.2 mA/cm^2, whereas on Ti–2% Ni alloy, under the same conditions, a potential of 1.9 V was reached and did not change during 20 h of polarization. Also, at a current density of 0.3 mA/cm^2, the value of E_a was not reached in 1 g/l NaCl solution during 100 h of polarization and no weight loss of the metal was noted.

The most striking difference is that between the polarization curve 4 obtained on the Ti–2% Ni alloy in chloride-alkali solution and all the other polarization curves. Above potential 0.6 V, the current density proliferated exponentially and at a potential value of 1.3 V, i.e., below the potential value of chlorine evolution, a current density as high as 70 mA/cm^2 was already reached. The current was almost completely spent on anodic oxygen evolution.

The high electrochemical activity of the alloy is, most probably, connected with two above-mentioned factors, which intensify each other. On the one hand, the presence of the intermetallic compound Ti$_2$Ni in the alloy increases the conductivity of the passive film; on the other hand, the barrier properties of the passive film formed in alkaline solutions are decreased.

Reduction of the oxygen overvoltage on the intermetallic compound Ti$_2$Ni may be connected with the presence of nickel in this compound. It was noted in Chapter 5 that nickel is characterized by a low overvoltage of oxygen evolution and, due to this property, it is used as the anode material in alkaline, chloride-free solutions of water electrolysis.

Thus, the alloying of titanium with nickel does not decrease the corrosion stability of the metal in chloride-containing media and increases the conductivity of the passive film that is formed on the metal surface. As a result, the electrochemical activity of the alloy drastically increases under anodic polarization in alkaline solutions.

The obtained results point to a promising future for the application of Ti–2% Ni alloys in the presence of external anodic currents, especially in chloride–alkali solutions. However, data [40] on the possible activation and intensive dissolution of this alloy under certain conditions after a long incubation period of time must be taken into account. Therefore, the final conclusions on the possibility of using Ti–2% Ni alloy in the conditions under consideration can only be drawn after long-time industrial testing.

REFERENCES

1. A. M. Sukhotin ed., Electrochemistry Handbook (in Russian), 1981, Khimiya, Leningrad, 486p.
2. Yu. S. Ruskol, Titanium Structural Alloys in Chemical Plants (in Russian), Khimia, Moscow, 1989, 286p.
3. A. M. Sukhotin and L. I. Tungusova, Zashchita Metallov, Vol. 7, No. 6, 1971, 654–659.
4. J. Lee, Corrosion (NACE) , 1981, Vol. 31, No. 8, 467–481.
5. N. D. Tomashov, Yu. S. Ruskol and G. A. Ayuyan, Zashchita Metallov, Vol. 10, No. 5, 1974, 515–519.
6. V. A. Levin, N. B. Kalinicheva and L. P. Chernikh, Khim.i Neftianoe Mashinostroenie, No. 3, 1976, 24–37.
7. L. M. Yakimenko, G. N. Kokhanov, I. E. Veselovskaya and R. V. Djagatzpanian, Khim. Promishlennost, No. 1, 1962, 43–47.
8. J. Dugdale and J. B. Cotton, Corrosion Science, Vol. 20, No. 4, 1964, 397–411.
9. F. A. Posey and E. G. Bohlmann, Desalination, 1967, Vol. 3, No. 3, 269–279.
10. M. W. Breiter, Electrocimica Acta, 1970, Vol. 15, No. 7, 1195–1200.
11. T. Koizumi and S. Furuya, Pitting Corrosion of Titanium in High Halide Solutions, In: Titanium Science and Technology, 1973, Vol. 4, New York – L., pp. 2383–2393.
12. Ya. M. Kolotirkin, Khim. Promishlennost', No. 9, 1963, 38–46.
13. V. V. Andreeva, Corrosion , 1964, Vol. 15, No. 2, 35–46.
14. O. Rudiger, W. Fisher and W. Knorr. Zeitschriff fur Metallhunde, 1956, Bd. 47, No. 8, s.s 599–604.
15. P. I. Krijanovsky, Izvestiya Vuzov, Khimiya i Khim. Tekhnologiya, Vol. 6, No. 3, 1963, 459–464.
16. I. L. Rozenfeld, Corrosion and Protection of Metals (in Russian), Metallurgiya, Moscow, 1970, 315p.
17. H. Hashe, Galvanotechnik, 1971, Bd. 62, No. 6, s.s 463–470.
18. N. A. Marchenko and G. G. Chernenko, Zashchita Metallov, Vol. 9, No. 1, 74–77.
19. F. Mazza, Corrosion, 1967, Vol. 23, No. 8, 223–230.
20. V. I. Ginzburg, L. G. Tetereva and A. F. Mazanko, Khim. Promishlennost, No. 10, 1978, 761–763.
21. Ya. M. Kolotirkin, Zashchita Metallov, Vol. 3, No. 2, 1967, 131–144.
22. A. J. Sedrics, J. A. Green and D. L. Novak, Corrosion, 1972, Vol. 28, No. 4, 137–142.
23. G. N. Kokhanov, Yu.V. Dobrov and L. A. Khanova, Electrokhimiya, Vol. 7, No. 7, 1971, 928–932.
24. G. N. Kokhanov, E. L. Krongauz and V. B. Busse-Machukas, Electrokhimiya, Vol. 14, No. 8, 1311, Dep. VINITI, 1978, No. 1257–1278 No. 8, стр. 1311. Деп. В ВИНИТИ апр. 1978, No. 1257–1278.
25. B. de Jelas, D. Sharket, M. Armand and R. Trikot, Transactions of the 3rd International Conference on Titanium,Vol. 2, VILS, Moscow, 1978, pp. 147–154.
26. J. B. Cotton and M. L. Green, Behavior of titanium welds in oxidizing acids. Proceedings of Third International Congress on Metals Corrosion, Vol. 1, 1969, Moscow, pp. 303–309.
27. S. M. Gurevich, Welding Metallurgy and Technology of Titanium and its Alloys (in Russian), Naukova Dumka, Kiev, 1979, 300p.

28. A. I. Gorshkov, Svarochnoe Proizvodstvo, No. 6, 1975, 54–56.
29. J. C. Griese, Corrosion, Vol. 24, No. 4, 1968, 96–109.
30. J. F. Soloviova, M. N. Fokin and V. A. Timonin, Zashchita Metallov, Vol. 6, No. 2, 1970, 204–206.
31. A. Takamura, Corrosion, 1967, Vol. 23, No. 10, 306–313.
32. Yu. S. Ruskol and I. Ya. Klinov, Zashchita Metallov, Vol. 4, No. 5, 1968, 495–499.
33. K. Shiobara and S. Morioka, J Jap. Inst. Metals, 1971, Vol. 35, No. 10, 980–984.
34. I. V. Riskin and A. V. Turkovskaya, Zashchita Metallov, Vol. 5, No. 4, 1969, 443–445.
35. V. I. Kazarin, N. D. Tomashov, V. S. Mikheev and B. A. Goncharenko, Zashchita Metallov, Vol. 12, No. 3, 1976, 268–271.
36. A. Z. Fedotova and E. N. Paleolog, Electrokhimia, Vol. 14, No. 6, 1978, 894–897.
37. A. G. Akimov, E. N. Paleolog, A. Z. Fedotova et al., Electrokhimia, Vol. 15, No. 7, 1979, 1089–1094.
38. N. D. Tomashov and Yu.M. Ivanov, Zashchita Metallov, Vol. 1, No. 1, 1965, 36–40.
39. T. P. Stepanova, V. V. Krasnoyarsky, N. D. Tomashov and I. P. Drujinina, Zashchita Metallov, Vol. 14, No. 2, 1978, 169–171.
40. Yu. S. Ruskol, T. A. Buraya, A. G. Parshin and E. I. Oginskaya, Zashchita Metallov, Vol. 15, No. 2, 1979, 213–215.

CHAPTER 7

HYDROGENATION AND CORROSION INVESTIGATIONS OF TITANIUM UNDER ATTACK BY AN EXTERNAL CATHODIC CURRENT

Contents

The purpose of the investigations discussed in this chapter consists, first of all, in the elucidation of the mechanism and features of titanium corrosion, discussed in previous chapters, under attack by cathodic leakage currents in chlorine-saturated NaCl solutions (electrolysis of NaCl solutions with a mercury cathode). In addition, it was of interest to study the conditions of titanium hydrogenation in the catholyte of nickel electrorefining (hydrogenation of titanium matrixes in the process of producing nickel cathode blanks). The results of these investigations were necessary for further develop the methods and means of titanium protection against corrosion and hydrogenation under attack by cathodic currents.

7.1. CONDITIONS OF HYDROGENATION AND CORROSION OF TITANIUM. METHODOLOGICAL FEATURES OF EXPERIMENTS

The tendency of titanium to undergo hydrogenation was known long ago [1]. The kinetic peculiarities of the hydrogenation process have been studied [2, 3], and it was found that the cathodic polarization of titanium in acid solutions leads to the growth of a hydride layer at a diminishing rate. At high cathodic current densities, titanium undergoes hydrogenation even in neutral solutions [3, 4].

Data on the corrosion of cathodically polarized titanium are more limited. It has been shown in Refs. [1, 5, 6] that the cathodic activation of titanium occurs in sulfuric and hydrochloric acids in the presence of air. Corrosion development in titanium under cathodic polarization in the presence of oxidizers such as nitric acid has also been reported [7]. According to the data [8], the corrosion rate of titanium was found to be constant in a certain cathodic potential range. The independence of the corrosion rate on the potential was explained by the chemical mechanism of titanium dissolution.

In studies of titanium corrosion behavior during cathodic polarization, very little attention was paid to the relationship between the hydrogenation and corrosion resistance of titanium. Even the limited data on the influence of titanium hydrogenation on its corrosion are ambiguous. It was noticed in Ref. [5] that in a preliminary study of titanium hydrogenation, its corrosion rate in sulfuric acid decreased. In later works [9, 10], data were given on the higher corrosion rate of hydrogenated titanium and titanium hydride with respect to the metal free of a superficial hydride layer. Cases of the metal losing its mechanical strength, of cracking and of mechanical destruction of the superficial layer as a result of the growth in volume of the hydrogenated layer (which causes internal stresses) were especially considered in Ref. [11].

Detailed investigations of the hydrogenation and corrosion of titanium under the conditions of attack by an external cathodic current were performed in a saturated chlorine solution of 260 g/l NaCl, i.e., the chloranolyte (technological solution used in electrolysis with a mercury cathode) and in a 1 g/l NaCl chlorine-saturated solution (a solution imitating the film of condensate that forms inside the wet chlorine header in diaphragm electrolysis).

The experiments were conducted in thermostated glass cells without stirring, with weak stirring by circulation (5 cm/s) in the cell circuit, or with intensive stirring and solution flow. In the last case, the solution was agitated by a magnetic stirrer rotating at 360 rpm. Under intensive stirring, the flow rate of the solution with respect to the working electrode was 30 cm/s. The stirring rod was placed on an auxiliary disc-shaped platinum electrode attached horizontally to the cell bottom. The volume of the through-flow cell was 5 ml. In the course of experiments, a chlorine-saturated solution of 260 g/l NaCl (pH 1.8–2) preheated to 90°C was fed at a rate of 1.2 ml/s from an auxiliary 5 l volume vessel to the through-flow cell. In the no-flow experiments, the solution was saturated by passing chlorine through the solution directly into the cell for 1h; during the experiments chlorine was passed above the solution. In the experiments with intensive stirring of the solution and flowing through the cell, a spade-shaped specimen was bent at 90° to the stem so that its working part was disposed horizontally, parallel to the auxiliary Pt electrode and at a distance of 3 mm above the stirring rod.

Electrochemical studies were carried out in the following regimes of polarization: galvanostatic, potentiostatic and with a stepwise potential variation (through 100 mV) in the range from 0 to −1.2 V. In the last case, the sample was polarized at each potential for 20 min. Preliminary experiments lasting for 6–10 h showed that the corrosion rate is stabilized in about 5–10 min.

The amount of titanium that passed into the solution was determined in some experiments (mainly in the galvanostatic regime) colorimetrically with diantipyrilmethane as a complexing agent. Primarily, the rate and kinetics of the titanium corrosion were measured by γ-radiometry of ^{47}Sc. It has been shown earlier that a foreign tracer (^{47}Sc) could be used for studying titanium corrosion [12, 13].

Hydrogen absorption (hydrogenation) was measured by the radioisotopic method, by using a tracer of radioactive tritium [14]. The titanium samples, after being polarized in the tritium-labeled solution (14–50 mCi/ml), were washed with distilled water and dissolved in 40% H_2SO_4 at 98°C. The evolving hydrogen was oxidized with CuO at 300°C. The amount of hydrogen absorption (hydrogenation) was determined by the radioactivity of the water obtained.

Relative errors in the corrosion and hydrogenation rates as measured in the above methods did not exceed 15%.

7.2. INVESTIGATIONS IN NON-STIRRED NaCl SOLUTIONS

The first step of the studies was performed under the conditions of galvanostatic polarization of the titanium samples. The object of this step was to evaluate the effects of the hydrogenation and corrosion of titanium during cathodic polarization.

The preliminary experiments showed that cathodic polarization in chlorine-saturated 1 g/l NaCl solution at a temperature of 98°C is accompanied by a transfer of titanium ions into the solution, whereas no titanium traces were detected without polarization. Hence, under these conditions, the titanium is subjected to corrosion connected with attack by the external cathodic current.

Alongside corrosion, hydrogenation of the titanium was observed. An anomalous character of the dependence of the hydrogenation on time under cathodic polarization is of interest (Figure 7.1). The dependence of the hydrogenation on time, mentioned earlier, obtained in acid solutions [3] are characterized by a gradual

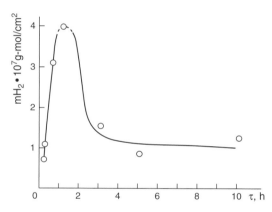

Figure 7.1 Titanium hydrogenation (mH_2) vs polarization time τ at cathodic current density of 20 mA/cm^2 in chlorine-saturated 1 g/l NaCl solution. Temperature, 98°C.

slowing down of the growth in the hydride layer. In the case under consideration, the hydrogen absorption increased rapidly during the first hour of polarization at a current density of $20 \, mA/cm^2$ and then sharply decreased, which may just be connected with the dissolution of the hydride layer. After 2–3 h of polarization, the hydrogenation remained constant, thereby indicating that the rates of formation and dissolution of the hydride layer had equalised.

The data obtained suggest that the corrosion of titanium passes through the stage of the formation of hydride and its subsequent transfer into solution. The hydride dissolution may be explained by its thermodynamic instability in the range of potentials of -0.1 to $-0.2 \, V$ that are established on titanium under the conditions considered [15, 16].

Corrosion studies of powders of hydrides of compositions TiH_2 and $TiH_{0.7}$ were performed, in order to verify this suggestion. Amounts of 0.6 g of each powder were tested at a temperature of 98°C in 1 g/l NaCl chlorine-free and chlorine-saturated solutions (prepared by bubbling chlorine through the solution throughout the time of testing). No traces of titanium were detected in the chlorine-free solution. In contrast, the titanium content was significant in the chlorine-saturated solution. The corrosion rate of the $TiH_{0.7}$ and TiH_2 powders, taking into account their specific surface areas were equal, was estimated at about 0.01 and 0.1 μm per 5 h of testing, respectively; i.e., the higher the hydrogen content, the higher the dissolution rate of the hydride. Thus, the results obtained prove that the hydrides are not stable in solutions containing dissolved chlorine.

It should be noted that testing the powders without polarization can be used only for a qualitative confirmation of the fact that the presence of dissolved chlorine, which raises the redox potential of solution, causes the hydride dissolution.

It was assumed that the diminishing hydrogenation is caused by the formation of the oxygen–chlorine compounds, chlorate and hypochlorite, in the near-electrode layer (close to the surface of the cathode) as a result of alkalization. Hypochlorite is formed in the reaction of alkali with the dissolved chlorine [17].

$$Cl_2 + 2NaOH = NaOCl + NaCl + H_2O \qquad (7.1)$$

Chlorite is the product of hypochlorite decomposition at elevated temperatures [18].

$$3ClO^- = ClO_3^- + 2Cl^-, \qquad (7.2)$$

It is also formed in the reaction of the dissolved chlorine with the hydroxyl ions in the near-electrode layer.

$$3Cl_2 + 6OH^- = ClO_3^- + 5Cl^- + 3H_2O \qquad (7.3)$$

The assumption regarding the influence of oxygen–chlorine compounds is corroborated by the data presented by Kuchinsky and Kochanov [19] in which a sharp decrease in the hydrogen absorption of cathodically polarized iron and tantalum was observed in alkaline NaCl solution in the presence of 0.5 g/l hypochlorite.

Wide potential variations are possible under galvanostatic polarization owing to the changes in the content of chlorine and oxygen–chlorine compounds in the solution. Therefore, the next steps involved investigations in a potentiostatic mode.

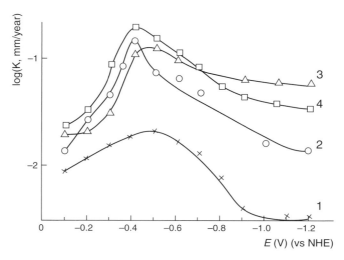

Figure 7.2 Titanium corrosion rate (K) vs polarization potential (E) in solutions at pH 2 and NaCl concentrations: 1 and 3 – 1 g/l; 2 and 4 – 260 g/l (1 and 2 – chlorine-saturated solutions; 3 and 4 – chlorine-free solutions).

The experiments performed in the 1 g/l NaCl chlorine-saturated solution have shown [20] that the dependence of the corrosion rate on the potential has a maximum at −0.5 V, the corrosion rate being 0.02 mm/year (Figure 7.2, curve 1). In chlorine-saturated 260 g/l NaCl solution, the corrosion rate of titanium is higher throughout the entire potential range and has a maximum at −0.4 V (0.14 mm/year) (curve 2). The increased corrosion rate in concentrated NaCl solution may be explained by the fact that the solubility of chlorine drops to almost 12% of that in 1 g/l NaCl solution [21]; consequently, the content of oxygen–chlorine compounds decreases.

There are, however, data [22] demonstrating that when the pH is constant, the corrosion rate of titanium rises with the increase in the NaCl concentration. To evaluate the effect of the salt concentration on the corrosion of titanium, the corrosion rate was studied in relation to the potential in acidified (pH 2) and de-aerated solutions of 1 and 260 g/l NaCl (Figure 7.2, curves 3 and 4). It was seen from these data that the corrosion rates in solutions of 1 and 260 g/l NaCl differ by no more than 1.5–2 times, whereas in chloride solutions saturated with chlorine, the corrosion rate in a concentrated NaCl solution is 6–8 times higher than that in a dilute solution along almost the entire potential range. This gives grounds to believe that the concentration of the oxide–chlorine compounds is the major factor determining the corrosion behavior of titanium in chloride solutions containing dissolved chlorine. This conclusion is also confirmed by the fact that the titanium corrosion decreases when comparing chloride solutions without chlorine (in which no oxygen–chlorine compounds are formed) to those saturated with chlorine. Thus, in 1 g/l NaCl solutions, the corrosion rate decreases by almost one order of magnitude (curves 1 and 3), and in 260 g/l NaCl it decreases by only 2–3 times (curves 2 and 4).

The data obtained point out, inter alia, that the primary cause of titanium hydrogenation and corrosion under cathodic polarization in chlorine-saturated

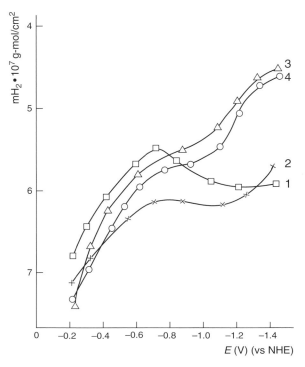

Figure 7.3   Titanium hydrogenation (mH$_2$) vs potential E in NaCl solutions: 1 and 3 – 1 g/l; 2 and 4 – 260 g/l. 1 and 2 – chlorine saturated; 3 and 4 – acidified (pH$_2$) chlorine-free solutions. Temperature, 98°C.

solutions is the reduction of their pH to values of 1.8–2, as a result of the hydrolysis of dissolved chlorine. In fact, experiments performed in the same range of potentials in nitrogen saturated solutions showed that the titanium corrosion rate, irrespective of the stirring conditions, was very low, no more than 0.001 mm/year [20]. Therefore, no noticeable corrosion takes place in neutral brine piping. However, acidification of the chlorine-free solution of chloride by hydrochloric acid, to pH values of 1.8–2 increases the corrosion rate by almost two orders of magnitude (Figure 7.2, curve 4) and increases the hydrogenation of the metal (Figure 7.3, curves 3 and 4).

The dependence of the hydrogenation on the potential in chlorine-saturated 1 g/l NaCl solution has a pronounced maximum at –0.7 V (Figure 7.3, curve 1). In chlorine-saturated 260 g/l NaCl solution, the hydrogenation is considerably lower at potentials more negative than −0.8 V, possibly due to the more intensive dissolution of the surface hydride layer in concentrated solutions (see Figure 7.2, curves 1 and 2). However, the hydrogenation increases anew at more negative potentials. In chlorine-free solutions, the hydrogenation increases in the entire potential range (Figure 7.3, curves 3, 4).

One of the factors that may lead to acceleration of the titanium corrosion in industrial conditions is fluctuations over a wide range of leakage currents at the areas of their attack on titanium constructions [10]. To evaluate the influence of these fluctuations the following experiments were conducted. The samples were alternately subjected to the action of −0.3 and 1.2 V in chlorine-saturated 1 g/l NaCl solution for 1 min at each potential up to the total experiment duration of 60 min. Under these conditions the corrosion rate increased by more than one order of magnitude, attaining an average of 0.5–0.6 mm/year and more. This result, along with the other data [10], points to another possible reason for the acceleration of the titanium corrosion.

Thus, a decrease in the pH, an increase in the NaCl concentration and current fluctuations are the factors that promote titanium corrosion by cathodic currents. With the increase in the NaCl concentration, the content of dissolved chlorine in solution decreases; consequently, the concentration of the formed oxygen–chlorine compounds (compounds inhibiting the hydrogenation and corrosion of titanium) diminishes which, in turn, leads to an increase in the hydrogenation and corrosion of titanium under attack by the cathodic currents.

7.3. INVESTIGATIONS UNDER CONDITIONS OF ELECTROLYTE STIRRING AND FLOWING

As the accumulation of oxygen–chlorine compounds near the electrode surface and in the electrolyte bulk depends on the stirring and solution flow, experiments were performed with due regard for these factors.

7.3.1. Relation between the corrosion rate of titanium and the accumulation of oxygen–chlorine compounds in the solution

The experiments were carried out in chlorine-saturated 260 g/l NaCl solution in which the corrosion rate is higher than in a dilute solution. The dependences of the corrosion rate on the titanium potential were studied under the following conditions (Figure 7.4): 1 – without stirring (curve 1); 2 – in a solution stirred by

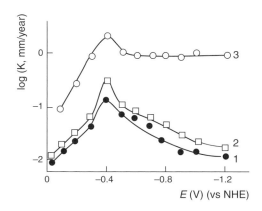

Figure 7.4 Corrosion rate of titanium (K) vs polarization potential (E) in chlorine-saturated 260 g/l NaCl solution: 1 – without stirring; 2 – stirred solution; and 3 – stirred flowing solution.

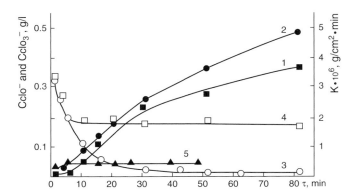

Figure 7.5 Kinetics of accumulation of oxygen–chloride compounds (1, 2 and 5) and corrosion losses on titanium (3 and 4) at potential $E = -0.4\,V$ in chlorine-saturated 260 g/l NaCl solution: $1 - ClO^-$; $2 - ClO^{3-}$; and $3 - (ClO^- + ClO^{3-})$. 1, 2 and 3 – stirred in a closed volume; 4 and 5 – stirred through-flowing solution. Temperature, 90°C.

circulation (5 cm/s) in a closed volume (curve 2); and 3 – in a stirred flowing solution (curve 3). All the curves have maxima at –0.4 V. The corrosion rate was higher by one order of magnitude in case 3 than it was in case 1 over the whole potential range and attained 2.12 mm/year at –0.4 V, i.e., it was comparable to the corrosion rate found under industrial conditions (about 6 mm/year).

The accumulation of hypochlorite and chlorate during the cathodic polarization of titanium was studied in a stirred, non-flowing solution, at –0.4 V (Figure 7.5, curves 1 and 2). Simultaneously, the corrosion rate was determined (curve 3). With the increased concentration of ClO^- and ClO^{3-} in the solution inside the cell, the corrosion rate of the titanium slowed sharply. During the first 30 min of the experiment, the concentration of hypochlorite and chlorate ions reached 0.23 and 0.26 g/l, respectively, and the corrosion rate decreased to $0.25 \times 10^{-6}\,g/cm^2$ (0.24 mm/year).

Since oxygen–chlorine compounds do not accumulate in a through-flowing solution (curve 5), the titanium corrosion at a potential of –0.4 V in this solution decreased only in the first five minutes and then remained constant ($1.8 \times 10^{-6}\,g/ cm^2$ min or 2.12 mm/year) (curve 4). For the first 5 min, the initial sections of curves 3 and 4 coincided because the hypochlorite and chlorate in the bulk of flowing and stagnant solutions had not accumulated in sufficient concentrations during this time interval (curves 1, 2 and 5).

Thus, the low corrosion rate in the stagnant solution is related to the accumulation of hypochlorite and chlorate in the solution.

The results obtained explain why neither corrosion nor hydrogenation of titanium by cathodic currents takes place in wet chlorine piping: oxygen–chlorine compounds accumulate in a thin condensate layer and inhibit both of these processes.

Figure 7.6 (curves 1 and 2) illustrates the relationships between the corrosion rate of titanium in a stirred flowing solution and the concentration of hypochlorite, introduced preliminarily into the chlorine-saturated 260 g/l NaCl solution. The

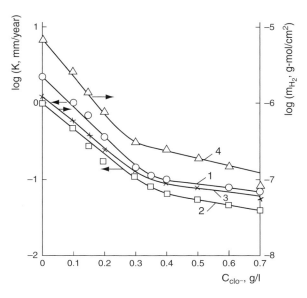

Figure 7.6 Corrosion rate of titanium (K) (plots 1 and 2) and hydrogenation (mH_2) (plots 3 and 4) vs hypochlorite concentration C_{ClO^-} in chlorine-saturated 260 g/l NaCl solution: 1 and 3 – (−0.4) V; 2 and 4 – (−1.0) V. Temperature, 90° C.

data confirm the role of oxygen–chlorine compounds in inhibiting the corrosion process, irrespective of the stirring and flow conditions, and agreeing well with the results given in Figure 7.5. Both the accumulation of hypochlorite during the polarization and the introduction of 0.3 g/l ClO⁻ into the solution, decrease the titanium corrosion rate by approximately one order of magnitude.

Introduction of chlorate into the 260 g/l NaCl solution also sharply decreases the corrosion rate of cathodically polarized titanium (Table 7.1).

Further, it will be shown that the revealed inhibiting effect of titanium hydrogenation and corrosion can be used to protect titanium structures against corrosion attack by external cathodic currents.

Table 7.1 Influence of sodium chlorate on corrosion rate (K) of cathodically polarized titanium in NaCl solutions

Potential (V)	Stirring	Concentrations (g/l)		K (mm/year)
		NaCl	NaClO$_3$	
			0	0.230
−0.4	Yes	260	0.5	0.076
			1.0	0.006
			0	0.021
−0.5	No	1	0.5	0.002
			1.0	0.001

7.3.2. Two corrosion mechanisms of hydrogenated titanium

The effect of the hypochlorite content on the corrosion rate was studied at potentials -0.4 and -1.0 V. These potential values were chosen because the first (corresponding to the maximum corrosion) lies in the range -0.2 to -0.6 V where the corrosion rate is potential-dependent, and the second is in a range where the corrosion rate is potential-invariant (Figure 7.4, curve 3). This difference may be due to diverse corrosion mechanisms. The effect of the hypochlorite content on the titanium corrosion and hydrogenation in stirred flowing solutions proved to be similar both at -0.4 and at -1.0 V (Figure 7.6). These data confirm the assumption that the corrosion proceeds through the stage of formation of the hydride; they also indicate that the decrease in the corrosion rate with the increasing concentration of hypochlorite is related to the decrease in the hydrogenation. Such behavior was observed in both the potential ranges under consideration.

Thus, the differences in the mechanisms of titanium corrosion are determined by the way in which titanium hydride is formed during the cathodic polarization and is transferred into the solution. This conclusion is supported by the fact that at $pH \leq 2.3$, titanium hydride is thermodynamically unstable in the potential range from -0.2 to -0.6 V, but is stable at higher cathodic potentials [15, 16]. Titanium corrosion under these conditions may result from mechanical fracture of the brittle surface layers of titanium hydride. This agrees with the above-mentioned data presented by Tomashov et al. [11]. They observed local failures of a titanium electrode with a scanning electron microscope and attributed them to chipping of the hydride layer formed at high cathodic potentials resulting from internal stresses that appear in this layer.

To verify that titanium is transferred into solution differently depending on the potential, hydrogenation and corrosion were tested side by side at -0.4 V (maximum corrosion) and -1.0 V (the region of potential-invariant corrosion rate). Titanium specimens (pre-radiated in a nuclear reactor) were polarized for 30 min at -0.4 and -1.0 V in a vigorously stirred, flowing 260 g/l NaCl solution saturated with chlorine and labeled with tritium. The solutions were filtered through pores no greater than 0.5 μm in diameter; the amount of titanium collected on the filter was determined by gamma activity. The filters were then washed with 2 l of distilled water, which completely removed the tritium-labeled water (as a control experiment showed). With the help of the above-mentioned technique [14], the hydrogen content was determined in the precipitate on the filter.

Only 3% of the total amount of titanium that passed into solution was detected on the filter after polarization at -0.4 V. On the other hand, after polarization at -1.0 V, this amount was equal to 46.3%. At -0.4 V, no tritium was found in the precipitate. The tritium content in the precipitate determined at -1.0 V suggests that the ratio of the hydrogen atoms to titanium atoms retained by the filter was close to 1. This proves that at -1.0 V, titanium passes into solution not in the ionic form, but as hydride particles chipped from the surface layer of the titanium. After chipping, some of the hydride particles may dissolve, since their potential must be shifted toward the positive region, i.e., to the range of thermodynamic instability of titanium hydride.

Table 7.2 Penetration of hydrogen into titanium specimens

No. of layers	Potential −0.4 V		Potential −1.0 V	
	Thickness of layers (μm)	Average atomic ratio H/Ti in the layer	Thickness of layers (μm)	Average atomic ratio H/Ti in the layer
1	0.66	0.12	0.56	0.98
2	0.56	0.10	0.62	0.92
3	0.93	0.088	1.23	0.86
4	0.23	0.046	0.73	0.32
5	1.64	0.021	0.62	0.11
6	1.89	0.09	2.17	0.068
7	2.19	0.01	2.38	0.043
8	–	–	3.31	0.008
9	–	–	1.78	0.003

In these experiments, the depth distribution of hydrogen in the titanium specimens after polarization was also studied. The specimens were rapidly dissolved in 40% H_2SO_4 during which the metal atoms passed into the solution, while hydrogen gas evolved simultaneously. The depth distribution of hydrogen in a sample was determined by repeating this procedure. Table 7.2 shows that for the specimens polarized at −0.4 V, the average composition of the 0.66-μm-thick surface layer was $TiO_{0.12}$. In the specimens that were polarized at −1.0 V, the hydrogen content was higher by one order of magnitude. The average composition of the 0.56 μm thick surface layer was TiH. Hydrogen had penetrated to a depth of 13.20 μm.

Thus, in the region of the potential where hydride is thermodynamically unstable, the corrosion of titanium is caused by the dissolution of hydride, whereas at more negative potentials it is caused by embrittlement of the surface and subsequent chipping of the hydride particles.

7.4. WELDING SEAMS

The data on the reduced corrosion resistance of the welding seams on titanium given in Chapter 6 were obtained in different media, taking into account the influence of anodic external currents. It was shown that welding seams sharply decrease the corrosion resistance of titanium under anodic polarization. Areas of attack, not only of anodic, but also of cathodic leakage currents, often coincide with the zones of welding seams on the metallic structure. This raises the practical importance of estimating the corrosion stability of welding seams on titanium under attack by external cathodic currents.

Cathodically polarized welded titanium specimens were preliminarily studied in 1 N H_2SO_4 (solution 1). The major part of the investigations was carried out in chlorine-saturated 260 g/l NaCl (solution 2) under stirring and flowing conditions and in chlorine-saturated 1 g/l NaCl solution (solution 3). The temperature of the solutions in all the experiments was 90°C. The corrosion was studied under potentiostatic polarization at −1.0 V on spade-shaped specimens joined together by a transvers seam produced by arc welding in argon. The hydrogenation was investigated on butt-welded titanium wire, 2 mm in diameter.

After polarization in solution 1 for 2.5 h and in solution 2 for 8 h, the welded zone was clearly revealed as a result of corrosion damage which occurred only in this zone (Figure 7.7). It is more clearly visible on a profilogram (Figure 7.8) obtained for the sample in solution 1, in which the corrosion of the welding seam was especially pronounced. Destruction of the metal is slower in the heat-affected zone than in the

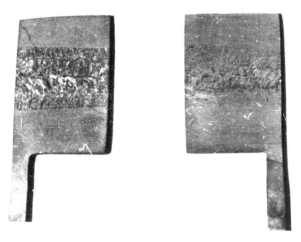

Figure 7.7 Welded specimens of titanium after polarization in chlorine–saturated 260 g/l NaCl (left) and in 1 N H_2SO_4 solutions. Temperature, 90°C.

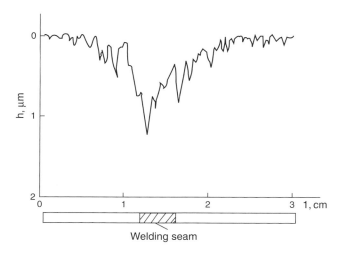

Figure 7.8 Profilogram of titanium welded sample after polarization in 1 N H_2SO_4 solution. Under the abscissa axis a seam is given from the welded section of the specimen used for recording the profilogram.

welded seam, but much more intensive than outside the zone of the welding influence. In solution 3 no corrosion was observed after polarization for 40 h. It should, however, be noted that some experiments in this solution, e.g., for 18 h at -1.8 V, also revealed a predominant failure of welding seams. In this case, the current density attained 0.5 A/cm^2 – an order of magnitude higher than that observed at -1.0 V.

Solution 3 simulates the chlorine-saturated condensate inside the piping for wet chlorine. No corrosion of the welding seams on titanium piping for wet chlorine was observed, since such high current values are not reached on these lines.

The results of the hydrogenation of welded samples of titanium wire agree with the above data on the corrosion of welded, spade-shaped titanium specimens. The highest hydrogenation in solutions 1 and 2 was observed at the welded seam zone (Figure 7.9); in the heat-affected zone it was noticeably reduced, and was the lowest outside of the zone. This was observed after the tests in both sulfuric and chloride solutions, although in chloride solution the hydrogenation was lower by approximately one order of magnitude.

The corrosion of welded seams and heat-affected zones at -1.0 V results from chipping of hydride particles from the surface and proceeds as described above. These zones suffer the predominant destruction as they are subjected to the greatest hydrogenation.

The results obtained indicate that at the sections of titanium structures that may be attacked by external currents of either anodic or cathodic directions it is most desirable to avoid the presence of welding seams.

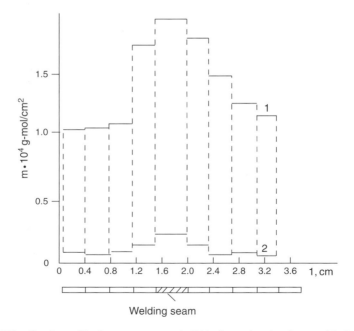

Figure 7.9 Distribution of hydrogen content (mH$_2$) along the titanium welded specimen in solutions: 1–1 N H$_2$SO$_4$; and 2 – chlorine saturated 260 g/l NaCl. Under the abscissa axis a seam is given at the welded wire specimen section.

7.5.   HYDROGENATION OF CATHODE BLANKS IN SOLUTION OF NICKEL ELECTROREFINING

In the process of nickel electrodepositing, the problem arose of hydrogenation of the titanium matrixes (used for producing cathodic blanks) [23]. As a result of hydrogenation, "joints" are formed on the surface of the matrixes which hinder the execution of stripping off the deposited blank. Moreover, hydrogenation leads to warping and cracking of the matrixes, which reduces their lifetime to less than 2 years.

7.5.1.   Electrochemical investigations of combined discharge of nickel and hydrogen ions on titanium

Hydrogenation is the result of a combined discharge of hydrogen and nickel ions under which both the cathodic deposit and the titanium matrix undergo hydrogenation. In this connection, investigations of the combined discharge kinetics of hydrogen and nickel ions, as well as of titanium hydrogenation, were carried out in the catholyte of nickel electrorefining.

Cathodic polarization curves on titanium were executed under galvanostatic (5 h long) and potentiodynamic (scanning rate, 0.72 V/h) conditions. Some of the experiments were performed in a solution free of Ni^{2+} ions (57.6 g/l NaCl + 2 g/l H_2SO_4), to determine the share of the current that was spent for the discharge of hydrogen ions. The temperature of all the solutions was 85°C.

The potentiodynamic cathodic polarization curves obtained in the above-mentioned solutions titanium and nickel (or a nickel coating) are shown in Figure 7.10. The reversible hydrogen electrode potential in these solutions, calculated from the Nernst equation, is equal to 0.12 V, and the potential of the nickel deposition is

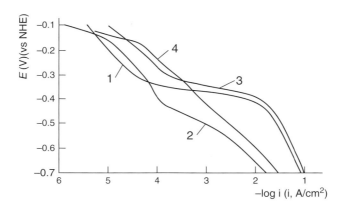

Figure 7.10   Cathodic polarization plots on titanium (1 and 2) and on nickel (3 and 4) in catholyte of nickel electrorefining (1 and 3) and in a solution of 57.6 g/l NaCl + 2 g/l H_2SO_4 (2 and 4). Temperature, 90°C.

Table 7.3 Hydrogen efficiency η and its evolution rates VH_2 at different current densities i under conditions of galvanostatic polarization

$i\,(mA/cm^2)$	$E\,(V)$	Potentiodynamic polarization		Galvanostatic polarization	
		$\eta\,(\%)$	VH_2 $(ml/cm^2\,h)$	$\eta\,(\%)$	VH_2 $(ml/cm^2\,h)$
1	−280	100	0.02	24.0	0.10
5	−340	43	0.03	7.0	0.15
20	−415	1	0.08	5.6	0.47

−0.25 V. In the solutions with and without Ni^{2+} ions in the range of potentials from the open-circuit titanium potential (−0.1 V) to −0.25 V, the only possible cathodic process is the discharge of hydrogen ions. Therefore, the initial sections of the cathodic curves 1 and 2 on titanium, and 3 and 4 on nickel, are close to one another. On nickel and nickel coating, the current densities are slightly higher due to the lower hydrogen overvoltage on nickel compared with that on titanium. At potentials more negative than −0.3 V, a bend, corresponding to the cathodic deposition of nickel, is noted on polarization curves 1 and 3 obtained in the solution containing Ni^{2+} ions. In the range of potentials from −0.3 to −0.4 V, the current density on the nickel is higher than that on titanium, owing to the already noted lower hydrogen overvoltage on nickel.

At potentials more negative than −0.4 V, the area of a limiting diffusion current for Ni^{2+} ions is observed. Curves 1 and 3 coincide in this area because the titanium surface, having been coated by deposited nickel, acquires the properties of a nickel surface. This was proved by the results of comparative experiments on specimens of nickel and nickel coated titanium. The cathodic curves obtained on these specimens coincided.

By comparing curve 2 with curves 1 and 3, and assuming that Ni^{2+} ions do not significantly influence the efficiency of the hydrogen evolution, it is possible to estimate this efficiency and the hydrogen evolution rate in the considered range of potentials (Table 7.3). Based on these results and the results of galvanostatic studies at current densities 1, 5 and 20 mA/cm^2, and taking into account the quantity of deposited nickel, it was found that the hydrogen evolution efficiency decreases when the cathodic current density increases. The higher hydrogen evolution rate under galvanostatic polarization can be explained by a reduction in the hydrogen overvoltage in the process of hydrogenation of the nickel coating [24].

7.5.2. Radiochemical investigations of titanium hydrogenation in the process of nickel deposition

To determine the kinetics of titanium hydrogenation under the conditions of nickel electrorefining, investigations were performed using radioactive tritium, which was introduced into the solution of nickel catholyte. The hydrogenation was determined in accordance with the procedure described above under galvanostatic polarization by a current density of $50\,mA/cm^2$. Two types of experiment were carried out:

1. Nickel was deposited on titanium for different periods of time, from 1 to 15 h. The maximum deposition time (15 h) represented the average time of nickel deposition on the matrix in industrial conditions. After the deposition process, the layer of the deposit was stripped mechanically (this operation is executed similarly when the deposit is stripped from the matrixes), and the titanium hydrogenation was determined.
2. Nickel was deposited on titanium for a period of 1 h then the deposit was stripped, and the one-hour deposition process was executed anew. Such cycles of deposition–stripping were carried out on different specimens 1–8 times; hydrogenation of the titanium specimen was measured after each stripping operation.

It is seen from Figure 7.11 that increasing the polarization duration from 1 to 15 h without stripping the deposit did not lead to a noticeable hydrogenation increase (curve 1), whereas at hourly deposit stripping, the hydrogenation is directly proportional to the number of operations (curve 2). These results prove the suggestion that hydrogenation takes place mainly in the initial period of nickel deposition, before the formation of a deposit layer that hinders the hydrogen penetration into the titanium.

All other things being equal, the possibility of metal hydrogenation by an external cathodic current in electrochemical plants, with metal deposition, is lower than that in electrochemical plants without metal deposition, since in the first case the major share of the current is spent for metal deposition. Along with this, it must be taken into account that the technological solutions of those plants often have a high acidity. Moreover, in the zones of attack by cathodic currents on titanium piping and equipment, the cathodic deposit is usually not formed: nickel is deposited in

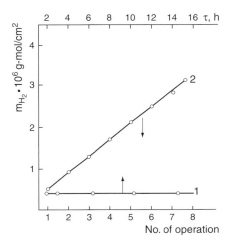

Figure 7.11   Dependence of titanium hydrogenation (mH$_2$) on: 1 – nickel deposition time (τ) without deposit stripping; 2 – No. of cycles of "deposition–stripping".

the form of dendrites, which are removed by the electrolyte flow. Therefore, despite the relatively low values of leakage currents (as compared to the values of the industrial current of nickel deposition), owing to their ability to concentrate on small surface areas, the danger of hydrogenation of titanium piping and equipment exists in nickel electrorefining plants.

REFERENCES

1. N. D. Tomashov and R. M. Altovsky, Corrosion and Protection of Titanium (in Russian), Mashinostroenie, Moscow, 1963, 168p.
2. N. D. Tomashov, V. N. Modestova, S. T. Glazunova et al., Investigation of Hydrogenation of α-titanium Alloys at Cathodic Polarization, Collected articles: "Corrosion of Metals and Alloys" (in Russian), Metallurgiya, Moscow, 1963, pp. 167–165.
3. Y. Y. Phyllips, P. Poole and L. L. Shreir, Corrosion Science, Vol. 12, No.11, 1972, 855–868; Vol. 14, No. 8, 1974, 533–542.
4. N. D. Tomashov, V. N. Modestova, E. S. Vlasova et al., Investigation of cathodic hydrogen absorption by titanium and cathodic attack of titanium. Proceeding of the Second International Congress Hydrogen in Metals, Vol. 3, 1977, pp. 1–8.
5. R. Otsuka, J. Electrohem, Soc., Vol. 26, No. 10, 1958, 191–194.
6. N. D. Tomashov, V. N. Modestova, L. A. Plavich and A. B. Averbukh, Investigation of Titanium Electrochemical Behavior, Collected articles: "Corrosion of Metals and Alloys" (in Russian), Metallurgizdat, Moscow, 1965, pp. 80–103.
7. A. F. Karasiov and A. I. Stabrovsky, Zashchita Metallov, Vol. 6, No. 3, 1970, 324–326.
8. F. Mansfeld and J. V. Kenкel, Corrosion Sci., 1976, Vol. 16, No. 9, 653–657.
9. N. D. Tomashov, V. N. Modestova, V. V. Usova and T. I. Strokopitova, Hydrogen Absorption by Titanium and its Alloys in Electrochemical Processes, Collected articles: "Metals Hydrogenation and Hydrogen Embrittlement Prevention" (in Russian), MDNTP, Moscow, 1973, pp. 148–155.
10. G. N. Trusov, E. P. Gochalieva and M. F. Fandeeva, Zashchita Metallov, Vol. 10, No. 1, 41–44.
11. N. D. Tomashov, V. N. Modestova and A. R. Yakubenko, Zashchita Metallov, Vol. 16, No. 2, 113–119.
12. M. A. Dembrovsky and N. N. Rodin, Zashchita Metallov, Vol. 3, No. 4, 1967, 517–519.
13. A. A. Uzbekov, G. G. Akalayev, I. V. Riskin and V. A. Likhobabin, Zavodskaya Laboratoriya, Vol. 38, No. 7, 1972, 816–818.
14. Ya. B. Scuratnik and V. B. Torshin, Zashchita Metallov, Vol. 16, No. 1, 46–49.
15. A. M. Sukhotin and L. I. Tungusova, Passivity of titanium and electrochemical behavior of its oxides and hydrides. Proceedings of 3rd International Conference on Titanium (in Russian)VILS, Moscow, 1978, pp. 43–47.
16. A. M. Sukhotin and L. I. Tungusova, Zashchita Metallov, Vol. 7, No. 6, 1970, 654–659.
17. H. Remy, Treatise on Inorganic Chemistry, Vol. 1, Elsevier, Amsterdam, 1956, 920p.
18. A. A. Furman, Oxidizing-Bleaching and Disinfecting Agents Containing Chlorine (in Russian), Khimiya, Moscow, 1976, 86p.
19. E. M. Kuchinsky and G. N. Kokhanov, Jurn. Fiz. Khimii, Vol. 36, No. 3, 1962, 480–488.
20. I. V. Riskin, V. B. Torshin, Ya. B. Skuratnik and M. A. Dembrovsky, Zashchita Metallov, Vol. 16, No. 2, 1980, 147–150.
21. M. I. Pasmanic, B. A. Sass-Tisovsky and L. M. Yakimenko, Producing of Chlorine and Caustic Soda, Handbook (in Russian), Khimiya, Moscow, 1968, 308p.
22. J. P. Frayret and R. Pointeau, Electrochimica Acta, Vol. 26, No. 12, 1981, 1783–1788.
23. A. I. Levin, Electrochemistry of Non-Ferrous metals (in Russian), Metallurgiya, Moscow, 1982, 256p.
24. L. I. Krishtalik, Electrokhimiya, Vol. 2, No. 5, 1966, 616–620.

CHAPTER 8

ESTIMATION OF CORROSION STABILITY OF STRUCTURES MADE OF PASSIVE METALS IN AGGRESSIVE MEDIA, IN THE FIELD OF AN EXTERNAL CURRENT

Contents

Data presented in the previous chapters have shown that an external current has to be considered as a major factor of environmental aggressiveness in relation to a metallic structure. At the same time, the initial state of the metal, its corrosion and electrochemical characteristics, as well as the composition and properties of the aggressive medium and the technological parameters and design peculiarities of the metallic structure must be taken into account when its corrosion stability is estimated in the field of an external current.

Of course, the criteria and methods of estimating the corrosion stability of a structure made of passive metal in the field of an external current must be based on empirical parameters known or accessible for measurement and control.

The criteria and methods that can be used to estimate the metal corrosion state of a structure and to predict the working reliability of the structure in the field of an external current are considered in this chapter.

8.1. ESTIMATION BASED ON THE POTENTIAL VALUE

When the corrosion of metallic structures under attack by stray currents was considered in Chapter 2, it was noted that the composition and properties of the environment are in a great part indefinite and may significantly change either in space (e.g., for large piping networks) or in time (e.g., at different seasons). Therefore, important parameters, such as the metal potential measured at one point, cannot be considered as reliable criteria for the estimation of the corrosion state of the metal. In this connection, other criteria such as the character of variations in the potential along a structure or at different periods of time are used. A great number of measurements are necessary to increase the reliability of these criteria. Nevertheless, results of these measurements on a carbon steel structure (the main structural material used for underground and underwater piping and structures) are not reliable enough. Due to the Taffel character of the anodic polarization curve on an active metal (see Figure 1.3), which is typical for carbon steel in soil and in water, the variations in the metal potential presence of external currents may not be clear enough to reveal its attack, and even less to estimate the degree of danger of this current.

8.1.1. Activation potential as an estimation criterion of the passive metal state in the field of an external current

In electrochemical plants, like in other chemical plants, the composition, temperature and other technological parameters of the aggressive solutions are specified in determined ranges, and corrosion–resistant structural materials are selected for these predetermined conditions, without taking into account the influence of external currents. In the majority of cases, the corrosion resistance of the selected metals is stipulated by their passive state. However, as already shown, under attack by an external current, the metal potential may shift from the value of the open circuit potential E_s to the value of the activation potential E_a.

It was noted in Chapter 3 that the direction of the leakage currents, and areas of their attack on piping and equipment, can be directly determined in an electrochemical plant from the equivalent scheme of the electrolysis electric circuit. By measuring metal potentials at the areas of leakage current attack and by comparing their values with the values of the open circuit potentials of the metal under similar conditions, it is possible to determine the metal state, since the characteristics of the metal and of the aggressive media are defined. For initially passive metals, the procedure is facilitated even further, owing to the significant value of the potential shift from the open circuit potential to the potential of the active state (see Chapters 1, 4–7): a significant jump in the potential is observed in the area where metal activation has occurred.

Under the influence of an anodic current, the stability of a metal structure is governed by the condition $E_{max} < E_{lim}$, where E_{max} is the maximum value of the potential of the metal in the structure and E_{lim} is the limiting permissible potential value under the conditions of attack by an external current.

When the considered area of the metallic structure is attacked by an external current of anodic direction, $E_{lim} = E_a$, and the stability condition of the structure can be written in the form of the inequality,

$$E_{max} < E_a, \tag{8.1a}$$

where E_a is the activation potential under the attack of an anodic external current.

When metal activation by an external cathodic current is possible, the stability condition can be expressed by the inequality,

$$E_{max} > E_{ca}, \tag{8.1b}$$

where E_{ca} is the activation potential under the attack of a cathodic external current.

Thus, the activation potential can be considered as a criterion that characterizes the state of the metal under the conditions of attack by an external current.

A sufficiently exact prediction of the metal corrosion state of a structure can be done only in the case where the potential distribution over the surface of the structure is known. This distribution depends on the shape and dimensions of the structure, on the polarization characteristics of the metal and on the size of the external current [1, 2]. Therefore, to determine the extreme potential values that may be reached in the structure, it is necessary, in each case, to solve the specific problems of the potential distribution over the metal surface of the structure, which is with a task with significant mathematical difficulties. A number of assumptions must be made to obtain the equations required for a solution by mathematical methods or with the help of a computer [3–8].

8.1.2. Types of structural elements in the form of tubes in the field of an external current

In estimating of the corrosion action of leakage currents in electrochemical plants, the greatest practical interest is attached to the design of tubes of various types. The technological pipelines for supplying and removing electrolytes are the most widespread type of structure to be found in the electrolysis zone along a set of electrolyzers. Moreover, structural elements such as connecting tubes of the pipelines and apparatus, nozzles, fittings and valves can, to a fair approximation, be regarded as tubular in shape. Therefore, finding the character of the potential distribution over the internal tube surface makes possible the solving of most of the practical problems connected with predicting the corrosion stability not only of piping, but also of structural elements of other kinds of equipment.

As the external current acts on the internal surface of a tube containing electrolyte, the field of this current is oriented along the tube in each of its sections, irrespective of its shape.

Tube-shaped structural elements that are found in practice can be divided into two major types: "bipolar" and "monopolar" tubes (Figure 8.1a and b).

Bipolar sections of tubes are those of finite length containing electrolyte and lying in the field of an external current and insulated from the metal of the rest of the

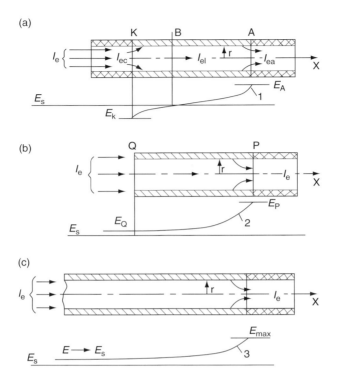

Figure 8.1   Types of tube-shaped structures (a–c) in the field of external current and metal potential distribution (1–3) along their internal surface: a, 1 – bipolar tube; b, 2 – monopolar tube; c, 3 – semi-infinite tube.

pipeline. One end of such a tube (K) is subjected to the action of a cathodic current (I_{ec}), and the other (A), to that of an anodic current (I_{ea}). With this approach, such a tube is similar to an extended bipolar electrode. Examples of structures which are similar to bipolar tube are sections of pipelines insulated from one another, metallic valves, T-pieces, slip rings, etc., insulated from the rest of the pipeline or mounted on a pipeline made of an electrical insulating material.

A tube directly connected to a communication line to which the external current enters can be considered as a monopolar tube. The most common example is the connecting tube of a metallic pipeline or an apparatus located in the field of an external current. In this case the current is practically fully concentrated at the end of this tube that is connected to the pipe of insulating material or from which the electrolyte flows out freely. The communication line is considered as a non-polarized current source with respect to the polarized end of the section of the monopolar tube.

As a limiting case of the above two examples, we shall also consider a model in the form of a semi-infinite tube (Figure 8.1c), with its end acted on by an external current. Lengthy sections can be considered, with a sufficiently high degree of accuracy, as semi-infinite tubes.

8.1.3. Potential and current distribution along structural elements in the form of tubes

The problem of the distribution of the potential over the internal surface of tubes for every kind of polarization curve was solved in Ref. [9]. For the case where the current density of the passive state of a metal is constant, the problem was solved in Ref. [10]. As opposed to the solutions obtained in Refs. [11–15], in Refs. [9, 10], the potential distribution was taken into account not only along the tube, but also along its radius. It was shown in work [12] that the potential distribution along the radius can be ignored only for tubes of a small diameter. Diameters of pipes in electrochemical plants reach 1 m and even more.

After double integration of the differential equation obtained in Ref. [9], the solution that characterizes the distribution of the potential of the metal along a tube was:

$$X(E) = \int_{E_1}^{E} \frac{\left[\frac{r}{2\rho} + \frac{r^2}{8} \frac{df(E)}{dE} \right] dE}{\sqrt{\frac{r^2}{8} f^2(E) + \frac{r}{\rho} \int_{E_1}^{E} f(E)\, dE + C_1}} + C_2 \qquad (8.2)$$

where r is the radius of the tube, ρ is the electrical resistivity of the electrolyte, $f(E)$ is the current density in a section X (polarization characteristic of the metal), E is the electrode potential of the metal at a given point of the surface of the structure, E_s is the stationary (open circuit) potential of the metal, and C_1 is the first integration constant which is determined from the boundary conditions, based on the current and potential values at the tube end. The second integration constant represents the coordinate of the section in which some potential E_1 is realized, when $X(E_1) = C_2$. If it is assumed that $C_2 = 0$, this means that the section in which $E = E_1$ is accepted as the origin of coordinates. It is reasonable to accept the open-circuit potential E_s as E_1, or the potential which is attained at the tube end.

The expression of the current distribution along the tube which is dependent on the potential is:

$$I_X(E) = 2\pi r \sqrt{\frac{r^2}{8} f^2(E) + \frac{r}{\rho} \int_{E_1}^{E} f(E)\, dE + C_1} \qquad (8.3)$$

8.1.3.1. Bipolar tube

The external current in the cross-sections K and A is the same ($I_A = I_K$) and is equal to the maximal leakage current I_e (Figure 8.1a). At cross-section B (between the ends K and A of the bipolar tube), which can be used as an origin of the abscissa ($X = 0$), the external current acting on the tube $I_e = 0$. Consequently, at this cross-section, the polarization current density (i.e., the external current density)

$f(E)=0$ and the stationary (open circuit) potential E_s is established on the metal. Any current which did not enter the tube wall flows in the electrolyte inside the tube through cross-section B, but it does not have any influence on the metal.

By substituting $I_x = I_0$ at $E = E_s$ and $f(E) = 0$ in (8.3), we obtain the constant of the first integration $C_1 = \left({}^{I_0} \big/_{2\pi r} \right)^2$. As cross-section B at which $E = E_s$ is taken as the origin of the coordinate X ($X = 0$), $C_2 = 0$, the expressions for the potential and current distributions along a bipolar tube of a finite length assume the form:

$$X(E) = \int_{E_s}^{E} \frac{\left[\dfrac{r}{2\rho} + \dfrac{r^2}{8} \dfrac{df(E)}{dE} \right] dE}{\sqrt{\dfrac{r^2}{8} f^2(E) + \dfrac{r}{\rho} \displaystyle\int_{E_s}^{E} f(E)\, dE + \left(\dfrac{I_0}{2\pi r} \right)^2}} \tag{8.4}$$

$$I_x(E) = 2\pi r \sqrt{\dfrac{r^2}{8} f^2(E) + \dfrac{r}{\rho} \int_{E_s}^{E} f(E)\, dE + \left(\dfrac{I_0}{2\pi r} \right)^2} \tag{8.5}$$

It is inferred from expression (8.4) that the current flowing inside the electrolyte in cross-section B, where $E = E_s$, $I_0 \neq 0$ at a finite length of the tube. In other words, the external current does not enter the tube wall entirely, a part of this current flows inside the electrolyte.

8.1.3.2. Monopolar tube

Along a tube of this type, the potential varies from the value E_p at the polarized end, P, to some value E_Q at the second end, Q (Figure 8.1b). It is most reasonable in this case to take as the abscissa origin ($X = 0$) the tube end Q at which $E = E_Q$, $I_x(E_Q) = 0$, and $E_1 = E_Q$. Under these conditions, the second constant $C_2 = 0$ and the first constant:

$$C_1 = -\frac{r^2}{8} f^2(E_Q) - \frac{r}{\rho} \int_{E_s}^{E_Q} f(E)\, dE$$

In this case the expressions for the potential and current distributions along the internal tube surface are as follows:

$$X(E) = \int_{E_Q}^{E} \frac{\left[\dfrac{r}{2\rho} + \dfrac{r^2}{8} \dfrac{df(E)}{dE} \right] dE}{\sqrt{\dfrac{r^2}{8} f^2(E) + \dfrac{r}{\rho} \displaystyle\int_{E_Q}^{E} f(E)\, dE - \dfrac{r^2}{8} f^2(E_Q)}} \tag{8.6}$$

$$I_X(E) = 2\pi r \sqrt{\frac{r^2}{8}f^2(E) + \frac{r}{\rho}\int_{E_Q}^{E} f(E)\,\mathrm{d}E - \frac{r^2}{8}f^2(E_Q)} \qquad (8.7)$$

8.1.3.3. Semi-infinite tube

In this type of tube, the total external current acts on the internal tube surface, i.e., $I_0 = 0$. It is impossible to obtain a solution for this tube by assuming $X = 0$ at $E = E_s$, because every point having a potential approximating to E_s will lie at infinity. In this case, the tube end at which E_{max} (maximal positive or negative potential value) has been realized is the most suitable for use as the abscissa origin. As mentioned above, the E_{max} can be accepted based on the activation potential value E_a under the considered conditions. At $E_1 = E_{max}$, the solutions for the semi-infinite tube are as follows:

$$X(E) = \int_{E_{max}}^{E} \frac{\left[\dfrac{r}{2\rho} + \dfrac{r^2}{8}\dfrac{\mathrm{d}f(E)}{\mathrm{d}E}\right]\mathrm{d}E}{\sqrt{\dfrac{r^2}{8}f^2(E) + \dfrac{r}{\rho}\displaystyle\int_{E_S}^{E} f(E)\,\mathrm{d}E}} \qquad (8.8)$$

$$I_X(E) = 2\pi r \sqrt{\frac{r^2}{8}f^2(E) + \frac{r}{\rho}\int_{E_S}^{E} f(E)\,\mathrm{d}E} \qquad (8.9)$$

8.1.4. Procedure for computation of the potential distribution along the internal tube surface with the help of a computer

Expressions (8.4), (8.6) and (8.8) enable a design of the potential distribution along the internal tube surface for different specific cases with the help of a computer. The design is executed in steps. ΔX_j – the distance between the cross-sections which have potential values of respectively, E_{j-1} and E_j, is calculated in every step.

$$\Delta X_j = \int_{E_{j-1}}^{E_j} \frac{\left[\dfrac{r}{2\rho} + \dfrac{r^2}{8}\dfrac{\mathrm{d}f(E)}{\mathrm{d}E}\right]\mathrm{d}E}{\sqrt{\dfrac{r^2}{8}f^2(E) + \dfrac{r}{\rho}\displaystyle\int_{E_S}^{E_{j-1}} f(E)\,\mathrm{d}E + \left(\dfrac{I_0}{2\pi r}\right)^2 + \dfrac{r}{\rho}\displaystyle\int_{E_{j-1}}^{E} f(E)\,\mathrm{d}E}} \qquad (8.10)$$

Function $f(E)$ is approximated by a set of functions of the type $A \exp(kE) + BE + C$. If the curve $f(E)$ is complicated, the area E is divided into a range of sections and the approximation is executed for each of them. Thereby, the integral that is disposed in the denominator can be solved analytically for each section in which the function $f(E)$ is expressed analytically.

If $\Delta E_j = E_j - E_{j-1}$ is small enough, it can be assumed that $A \exp(kE) \approx 1 - kE$ at the interval ΔE_j, and $\frac{df(E)}{dE} = \text{const}$, at the required degree of accuracy which can be determined by the computer program (on the basis of the ΔE_j value). After this simplification, expression (8.10) becomes

$$\Delta X_j = T \ln \frac{Tf(E_j) + \sqrt{\frac{r^2}{8} f^2(E_j) + \frac{r}{\rho} \int_{E_S}^{E_j} f(E)\, dE + S}}{Tf(E_{j-1}) + \sqrt{\frac{r^2}{8} f^2(E_{j-1}) + \frac{r}{\rho} \int_{E_S}^{E_{j-1}} f(E)\, dE + S}}, \quad (8.11)$$

where

$$T = \sqrt{\frac{r^2}{8} + \frac{r}{2\rho \dfrac{df(E)}{dE}}}, \quad S = \left(\frac{I_0}{2\pi r}\right)^2$$

In the process of calculation, the computer summarizes ΔX_j and produces results in the form of the dependence of X_j on E_j.

Thus, when the empirical data (polarization characteristics $f(E)$ and open circuit potential value of the metal E_s) and data on the electrolyte resistivity and tube diameter are given, it is possible to compute the potential distribution along the internal tube surface and to verify whether the condition (8.1) is met.

8.2. ESTIMATION BASED ON THE EXTERNAL CURRENT VALUE

Measurements of the metal potential inside the piping and equipment, especially under working conditions, present many technical difficulties and sometimes are just impossible. On the other hand, as shown above, measurements of external currents can be executed by accessible means. Therefore, an estimation of the corrosion stability of a metallic structure in the field of an external current based on the data of the current values is most feasible.

Based on this precondition and using the obtained equations of potential and current distributions inside the different types of tubes, criteria were established for the estimation of the stability of a structure when the values of the external currents are known.

8.2.1.   Estimation based on the external current value for tube-shaped structural elements

As mentioned above, the establishment of these criteria was based on the statement that the corrosion stability of a tube-shaped structure is determined by its ability to withstand the action of an external current, i.e., to retain its metal potential within the boundaries of its passive field which is limited by the metal activation potential. This means that the most extreme (i.e., maximal positive or negative) metal potential that is attained at the tube end will be below the activation potential E_a. Thus, the expression (8.1) remains in this case a major condition of the stability, and it is necessary only to find the relationship between the external current I_e and the maximal potential E_{max}, which is reached at the tube end.

It should be noted that in some cases the corrosion state of a metallic structure in the field of an external current can be preliminarily estimated without carrying out special calculations, but only by using the data on the values of the external current and the observation of the polarization curve in the passive field. For example, when the current density i_p in the passive field is small, in the order of magnitude of $10^{-5}-10^{-4}\,\mathrm{mA/cm^2}$ (titanium in neutral NaCl solutions), and the value of external anodic current measured on the structure is high, in the order of magnitude of amperes or tens of amperes, the probability of anodic activation of the metal in the areas of the external current attack is high. It is obvious that such an estimation has only a qualitative character.

In Ref. [16], which can be considered a logical continuation of Refs. [9, 10], criteria and methods of estimating the corrosion stability are proposed for the three above-mentioned types of metallic tubes and for every type of polarization curve.

8.2.2.1.   Semi-infinite tube

By using Equation (8.9), the relationship can be established between the external current I_e and the potential E_{max} at the end of a semi-infinite tube:

$$I_e = I(E_{max}) = 2\pi r \sqrt{\frac{r^2}{8}f^2(E_{max}) + \frac{r}{\rho}\int_{E_S}^{E_{max}} f(E)\,\mathrm{d}E} \qquad (8.12)$$

The limiting permissible current size I_{lim} conforms to the case when the limiting permissible potentials $E_{lim} < E_a$ or $E_{lim} > E_{ca}$ are established at the tube end under the action, respectively, of an anodic or cathodic external current:

$$I_{lim} = 2\pi r \sqrt{\frac{r^2}{8}f^2(E_{lim}) + \frac{r}{\rho}\int_{E_S}^{E_{lim}} f(E)\,\mathrm{d}E} \qquad (8.13)$$

Expressions (8.12) and (8.13), which give the modules of I_e and I_{lim}, i.e., $I_e \geq 0$ and $I_{lim} \geq 0$, are valid for both anodic and cathodic directions of external currents, and they enable the determination of the criteria of the stability of the tube metal, into which the external current value is included.

As the ratio I_e/I_{lim} is accepted as a criterion, the condition of the metal stability of the semi-infinite tube can be formulated as:

$$\frac{I_e}{I_{lim}} \leq 1 \qquad (8.14)$$

For a semi-infinite tube, the condition for stability (8.14) is equivalent to conditions (8.1a) and (8.1b). In contrast with condition (8.1), in which it is difficult to determine E_{max}, in (8.14) both quantities can be easily found. I_{lim} is calculated analytically, since the function $f(E)$ can be expressed in an analytical form. Polarization curves $i_p = f(E)$ for passive metals in different ranges of potentials can be represented by the exponential (Taffel) dependence or by the dependence $i_p = const$. The external current I_e is determined by measurement or calculation. Thus, the estimation of the stability based on the condition (8.14) is more accessible and more applicable than that based on condition (8.1).

It must be noted that the equivalence of conditions (8.1) and (8.14) is intrinsic, especially for a semi-infinite tube. Therefore, this type of a tube must be considered as a model which can used for the investigation of other important for types of tubes encountered in practice.

8.2.2.2.   Bipolar tube
For a bipolar tube the relationship between the external current and the potential at the end in question can be written, in accordance with expression (8.5), as follows:

$$I_e = I(E_{max}) = 2\pi r \sqrt{\frac{r^2}{8} f^2(E_{max}) + \frac{r}{\rho} \int_{E_S}^{E_{max}} f(E)\, dE + \left(\frac{I_0}{2\pi r}\right)^2} \qquad (8.15)$$

The external current will have the limiting permissible (maximal) value when $E_{max} = E_{lim}$. Hence, taking into account expression (8.13), we can obtain the condition for the stability of a bipolar tube which relates to the size of the external current I_e with the size of the limiting permissible current I_{lim}:

$$\frac{I_e}{\sqrt{I_{lim}^2 + I_0^2}} \leq 1 \qquad (8.16)$$

Thus, I_{lim} which is obtained for a semi-infinite tube represents a part of the criterion for the stability of a bipolar tube. From Equation (8.16) it directly follows as a sufficient condition for the stability of the bipolar tube, which coincides with condition (8.14). If (8.14) is satisfied, there is no need for further analysis. But if

(8.14) is not satisfied, the limiting permissible length of the bipolar tube L_{lim}^{b} can be found for which condition (8.16) will be satisfied.

Taking into account Equation (8.8) for the distribution of the potential along the internal tube surface, and also condition (8.16), it is possible to obtain an expression for the size of the limiting permissible length of the bipolar tube

$$L_{\text{lim}}^{b} = \int_{E_{P}}^{E_{\text{lim}}} \frac{\left[\frac{r}{2\rho} + \frac{r^2}{8} \frac{df(E)}{dE} \right] dE}{\sqrt{\frac{r^2}{8} f^2(E_{\text{max}}) + \frac{r}{\rho} \int_{E_S}^{E} f(E) dE + \frac{I_e^2 - I_{\text{lim}}^2}{4\pi^2 r^2}}}$$ (8.17)

Here E_{lim} is the limiting permissible potential at the "vulnerable" end. E_J is the potential at the conjugate end, found from (8.15) with I_0 calculated from condition (8.16) at the "vulnerable end: $I_0 = \sqrt{I_e^2 - I_{\text{lim}}^2}$.

In expression (8.17) it is assumed that $E_J < E_{\text{lim}}$; if $E_J > E_{\text{lim}}$, then it is more convenient to reverse the limits of the integration.

Expression (8.17) can be integrated on a computer in steps.

Using the value I_{lim} once found, it is possible to obtain a simple condition for the stability of a bipolar tube of length L:

$$\frac{L}{L_{\text{lim}}^{b}} \leq 1$$ (8.18)

In fact, the shorter the bipolar tube, the greater is the portion of the external current passing through the electrolyte and not acting on the metal, and the further are the potentials shifted at the ends of the tube from E_{lim} into the safe region.

In estimating L_{lim}^{b} in practice, simplifications are possible. If the polarizability at the "vulnerable" end is much lower than at the conjugate end, as in the case in which $I_J = \text{const}$, then the length of the conjugate end can be neglected. Equation (8.17) is integrated from E_s to E_{lim}, and there is no need to calculate E_J.

8.2.2.3. Monopolar tube

For a monopolar tube, the relationship between the external current and the potentials at the ends can be obtained from expression (8.7):

$$I_e = I(E_{\text{max}}) = 2\pi r \sqrt{\frac{r^2}{8} f^2(E_{\text{max}}) + \frac{r}{\rho} \int_{E_S}^{E_{\text{max}}} f(E)\, dE - \frac{r^2}{8} f^2(E_Q) - \frac{r}{\rho} \int_{E_S}^{E_Q} f(E)\, dE,}$$

(8.19)

where E_{max} and E_Q are, respectively, the potentials at the more strongly polarized end of the tube and at the opposite end, respectively.

As in the preceding case, the external current will have its limiting permissible value when $E_{max} = E_{lim}$. Hence we can obtain the condition of stability for a monopolar tube, relating the external current I_e with the limiting permissible current I_{lim} for a semi-infinite tube

$$\frac{I_e}{\sqrt{I_{lim}^2 - I_Q^2}} \leq 1 \qquad (8.20)$$

where

$$I_Q = 2\pi r \sqrt{\frac{r^2}{8} f^2(E_Q) + \frac{r}{\rho} \int_{E_S}^{E_Q} f(E) \, dE} \qquad (8.21)$$

i.e., I_Q is the external current for a semi-infinite tube at the end of which the potential is equal to E_Q.

From (8.20) it follows that condition (8.14) for semi-infinite tube is necessary but not sufficient for a monopolar tube. If it is not satisfied, it becomes necessary to use a metal with other characteristics or to take special protection measures. If (8.14) is satisfied, such a value for the length of the monopolar tube can be found, above which the condition of stability (8.20) will be satisfied.

Using the expression for the distribution of the potential along the monopolar tube (8.6) and substituting $I_Q = \sqrt{I_{lim}^2 - I_e^2}$, we get:

$$L_{lim}^m = \int_{E_Q}^{E_{lim}} \frac{\left[\frac{r}{2\rho} + \frac{r^2}{8} \frac{df(E)}{dE} \right] dE}{\sqrt{\frac{r^2}{8} f^2(E_{max}) + \frac{r}{\rho} \int_{E_S}^{E} f(E) dE + \frac{I_{lim}^2 - I_e^2}{4\pi^2 r^2}}} \qquad (8.22)$$

Expression (8.22) can be integrated in steps, similarly to that for a bipolar tube.

Using the value of L_{lim}^m, it is possible to obtain the condition for the stability of a monopolar tube of length L:

$$\frac{L}{L_{lim}^m} \geq 1 \qquad (8.23)$$

As the tube gets longer, the external current to it is distributed over a larger area of the internal surface, and the potential E_{max} is less shifted to the side of potential E_{lim}.

8.3. PRACTICAL STEPS AND EXAMPLES OF CORROSION STABILITY ESTIMATION OF STRUCTURES IN THE FORM OF TUBES OF DIFFERENT TYPES

The corrosion resistance of a metal structure is estimated in accordance with the above, in the following sequence.

1. Calculate I_{lim} for a semi-infinite tube and determine I_e/I_{lim}.
2. Determine whether the tube-shaped structure is bipolar or monopolar.
3a. In the case of bipolar tube for which the condition (8.14) is satisfied for the more "vulnerable" end, the estimation of the stability is over: the structure will remain stable regardless of its length. Otherwise, L_{lim}^b is calculated with the aid of Equation (8.17) and condition (8.17) is verified. If it is satisfied, the structure will remain stable for a given maximum permissible length. If (8.18) is not satisfied, the structure can be made stable by changing its length, by using materials with different electrochemical characteristics, by taking special measures to reduce external currents or by applying special measures of protection from electrocorrosion. These measures can be adopted independently of one another or in various combinations according to the particular conditions.
3b. If (8.14) is not satisfied for the case of a monopolar tube, protection can be obtained by taking the measures indicated in step 3a, apart from those involving changes in the length of the tube. If condition (8.14) is satisfied, then L_{lim}^m is calculated from Equation (8.22), and the stability condition (8.23) is tested. If it is satisfied, the estimation is over. Otherwise, stability must be obtained by taking the measures given in step 3a, except for those involving changes in the length of the tube.

For changing of the diameter as well as the length of the tube, the estimation of stability begins with step 1.

Let us estimate some examples of the practical application of the above-considered results in certain electrochemical plants.

Example 1. Estimation of possibility of using connecting tubes (nozzles) of titanium in titanium brine and chloranolyte pipelines of chlor-alkali electrolysis plants. The diameter of the nozzles is 50 mm.

It was shown above that the leakage currents in these pipelines reach 2 A. The polarization curve for these conditions in the range of potentials from E_s to E_a (Figure 8.2, curve 1) can be represented with a fair degree of accuracy in the form of two straight lines (Table 8.1, No. 1). Calculation revealed that $I_e/I_{lim} > 1$. Consequently, the nozzle, which we shall regard as a section of a monopolar tube, will fail in any length unless special protection is provided.

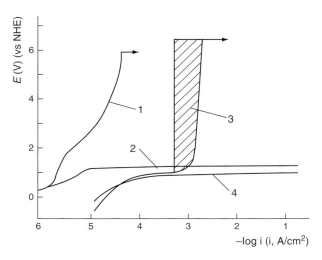

Figure 8.2 Anodic polarization curves on titanium (1 and 3), stainless steel 18–10 (2) and Ti–2% Ni (4). For conditions (see Table 8.1 under the same numbers).

Example 2. Sodium perborate electrolysis plant.

It was noted above that the headers for feeding solutions to the electrolyzers of these plants are made of stainless steel 18–10 and the leakage currents at the connecting tubes (nozzles) reach 1.6 A (Table 8.1, No. 2).

In the potential range from $E_s \approx 0$ to $E = 1.05$ V, the polarization curve (Figure 8.2, curve 2) can be represented as a straight line (Table 8.1, No. 2). At higher potentials the current rises exponentially with the potential.

The above considered tests on steel under anodic potentiostatic polarization revealed that up to a potential of 1.8 V the corrosion rate in the given medium does not exceed 0.1 g/m^{2}h and rises only at 1.9 V to 0.7–1.4 g/m^{2}h. To increase the reliability of the calculation, a lower value of $E_{lim} = 1.5$ V was adopted. Nevertheless, as seen from Table 8.1, $I_e/I_{lim} \ll 1$, i.e., a nozzle of steel 18–10 is able to remain stable in these conditions.

Examples 1 and 2 show that an important role is played by the polarization characteristics in choosing metallic materials for structures acted upon by external currents. Although the leakage currents are comparable, in the former case the nozzle is unstable, but in the latter case it is stable, despite the fact that the E_{lim} is much higher in the former case.

Example 3. Estimation of the stability of metallic overflow units for alkali liquor mounted on diaphragm electrolyzers in chlor–alkali electrolysis plants.

The leakage currents in the streams of electrolyte flowing out of these structures reach 2 A (Table 8.1, Nos. 3 and 4). The structures are in metallic contact with the steel cases of the electrolyzers (a source of leakage currents) and therefore can be regarded as of the monopolar tube type.

Table 8.1 Data for estimating corrosion resistance of tube-shaped metal structures (radius $r = 2.5$ cm) under the action of leakage currents

No.	Plant	Metal	Media characteristics					Electrochemical characteristics		I_e (A)	I_{lim} (A)
			Composition (g/l)	Temperature (°C)	ρ (Ωcm)	E_s (V)	E_{lim} (V)	Interval E(V)	Polarization characteristics $i = f(E)$		
1	Chlor-alkali	Titanium	300 NaCl	80	1.6	0	5.8*	0–1.6 1.6–5.8	$i = 0.25 \times 10^{-5}E$ $i = (3.8E{-}2.1) \times 10^{-6}$	2	0.13
2	Na perborate	Steel 18–10	150 Na_2CO_3 + 25 $NaHCO_3$	15	10.2	0	1.5**	0–1.05 1.05–1.5	$i = (0.037 + 0.92\,E) \times 10^{-5}$ $i = 0.32 \times 10^{-16} e^{25,\,2E}$	1.6	11.7
3	Chlor-alkali	Titanium	120 NaOH + 200 NaCl	90	1.1	−0.9	6.5*	−0.9 to 1.6 1.6–6.5	$i = (2E + 1.8) \times 10^{-4}$ $i_{min} = 5 \times 10^{-4} = \text{const}$	2	1.4
4	Chlor-alkali	Alloy Ti-2% Ni	120 NaOH + 200 NaCl	90	1.1	−1.0	1.5**	−1.0 to 0.6 0.6–1.5	$i = 2 \times 10^{-5} = \text{const}$ $i = 0.2 \times 10^{-8} e^{15,\,4E}$	2	300

* E_{lim} is taken to be equal to pitting potential E_a.
** Value of E_{lim} is underestimated.

For overflows of titanium, the polarization curve can be represented in the form of two straight lines (Figure 8.2, curve 3), and between $E = 1.6$ V and $E_{lim} = E_a = 6.5$ V, the current is practically independent of the potential.

As seen from Table 8.2, for titanium $I_e/I_{lim} > 1$, so that titanium overflow units will not be stable when acted upon by leakage currents; this is confirmed by the above-considered experience of their operation.

Let us consider the possibility of protecting them by insulating them from the cases of the baths. In this case, the overflow unit will be regarded as a bipolar tube; its maximum permissible length can be calculated using Equation (8.17). Calculations revealed that $L_{lim}^b = 68$ cm. In practice, the length of an overflow unit is $L = 100$ cm, i.e., $L/L_{lim}^b > 1$, and consequently insulation from the case does not afford protection. It is also undesirable, because it can lead to failure of the electrolyzer case due to leakage current attack in the insulation section.

Let us assess the possibility of replacing the titanium by the alloy Ti–2% Ni. The polarization characteristic of this alloy is quite different from that of titanium (Figure 8.2, curve 4). Above the potential value of 0.6 V, a Taffel dependence characterizing oxygen evolution takes place and the activation potential cannot be reached even in laboratory tests. As a deliberately underestimated value of $E_{lim} = 1.5$ V was accepted, it was found that $I_e/I_{lim} << 1$ (see Table 8.2) and correspondingly $L/L_{lim} >> 1$. Consequently, under these conditions the overflow units of alloy Ti–2% Ni will retain their corrosion resistance.

The given examples show that the developed estimation principles provide quite a lot of information on the corrosion stability of tube-shaped metallic structures in the field of an external current, and enable the definition of measures that are necessary for protection from attack by these currents in electrochemical plants of different types.

8.4. SIGNIFICANCE OF THE ELECTROCHEMICAL CHARACTERISTICS OF PASSIVE METALS FOR THE ESTIMATION AND PROVISION OF CORROSION STABILITY OF METAL STRUCTURES

The examples given in the previous section show that polarization characteristics and limiting permissible potential values determine, in many cases, the corrosion stability of a metal in the field of an external current. Some additional explanations are necessary in connection with these electrochemical characteristics of metals.

The stability condition $I_e/I_{lim} < 1$, which is equivalent to condition (8.1), means that the maximal possible density of the external anodic current, which is concentrated in the areas of its action on the metal, is lower than the anodic current density of the metal in a passive state. In most cases this is possible only when the external current is spent on the oxidation of the solution components on the passive metal surface at a potential value E_{ox}, which is lower than the activation potential E_a in the conditions under consideration. From this, it is inferred that the corrosion stability of a metallic

Table 8.2 Examples of metal-aggressive environment systems with various $\Delta = E_{ox} - E_a$ relationships

No.	Δ	Metal	Environment	Temparature (°C)
1	≥ 0	Carbon steel	120 g/l NaOH + 200 g/l NaCl; 1 N NaOH	90
		Steel 18-10	Electrolyte of copper electrorefining	60
2	< 0	Ti, Ti–0.2% Pd, Ti–5% Ta	NaCl solutions	80 – 100
3	$\ll 0$	Ti	Electrolytes of nickel and copper electrorefining (sulfate and chloride-sulfate solutions)	60 – 85
4	< 0	Steel 18-10	Electrolyte of perborate production	15
		Ti–2% Ni	120 g/l NaOH + 200 g/l NaCl	90
		Ni	Up to 50% NaOH	110

structure is mainly determined by the character of the polarization curve, since the ratio I_e/I_{lim} and the relationship between E_a and E_{ox} are determined by this curve.

Similar reasoning is possible for a cathodic current action on the metallic structure. In this case, the cathodic activation potential E_{ca} has to be more negative than the reduction potential E_{red} of some of the solution components, and the external cathodic current will be spent on the reduction of these components.

Based on the results of the executed investigations, the systems of metals in aggressive media can be divided into four types, in accordance with the indicated electrochemical characteristics (Figure 8.3 and Table 8.2). Each of the numbered curves in Figure 8.3 and the same numbered position in Table 8.2 relate to the correspondingly numbered type of the four types considered. Metal characteristics under the action of an external anodic current, which is usually more dangerous than a cathodic current, are considered.

For facilitating the analysis, it is worthwhile introducing a quantity that characterizes the relationship between the potentials of the oxidation of the solution components and the activation of metals $\Delta = E_{ox} - E_a$. If $\Delta \geq 0$, the metal activates at a potential which is more negative than the oxidation potential or in the region of this potential (Figure 8.3, curve 1 and Table 8.2, No. 1). Activation of the stainless steels in the copper electro-refining electrolyte occurs at the potential that is more negative than the potential of oxygen evolution ($\Delta > 0$). The activation potential and oxygen evolution potential on carbon steel in alkaline and chloride-alkaline media are close to one other ($\Delta = 0$).

A metal with a polarization characteristic corresponding to curve 2 in Figure 8.3 (Table 8.2, No. 2) does not activate when the E_{ox} value is reached ($\Delta < 0$). However, as a result of an oxide film barrier forming on the metal surface, only a small share of the external current is spent on such a metal for oxidation of the solution components at the potentials above E_{ox}. In the process of the oxide film growth under the action of the external anodic current, the metal potential continues to shift to the positive side and reaches the value of E_a.

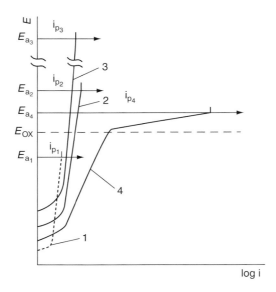

Figure 8.3 Typical anodic polarization curves for passive metals (explanations are given in the text).

Curves 2 and 3 differ from one another by a great quantitative difference between the activation potential values E_a, related to the difference in the activation mechanisms of these metals. It will be shown in Chapter 9 that this difference is significant when sectionalization of the piping is carried out in the field of the external current (Section 9.3).

The magnitudes of the current density in a passive state i_p of the metals of 2-d and 3-d types are usually small, owing to the barrier film formation. Thus, condition $\Delta < 0$ is necessary, but not sufficient, for preventing the potential shift of the metal to E_a under the action of the external current. It is also necessary that the magnitude of the current density of the oxidation of the solution components must prevail over the magnitude of the external current density below potential E_a. Type 4 satisfies this requirement. When potential E_{ox} is attained in such a system, no barrier film is formed and the external current is spent on oxidation of the solution components, but not on the metal activation.

In the example of type 4 given in Table 8.2, the kinetics of the oxidation of the solution components has a Taffel dependence. This means that the surface of the passive metal possesses a high electrochemical activity with respect to the oxidation of the solution components. As follows from a comparison of curves 2 and 4, just owing to this character of polarization of curve 4, the metal retains its corrosion stability under the action of the external current in the conditions under considera-tion. Despite the high values of the external current, the criterion of corrosion stability is $I_e/I_{lim} < 1$. It is seen from Example 3 of the previous section that the corrosion stability of a titanium structure was attained only by substitution of the titanium having a polarization characteristic of type 2, by the alloy Ti–2% Ni, which has a polarization characteristic of type 4 (Figure 8.3).

Electrochemical characteristics that are analogous to the ones considered above can be used for estimation of the stability of metal structures under attack by external cathodic currents, although the differences in the mechanisms of cathodic and anodic metal activation have to be taken into account. Cathodic activation of metal, if it occurs at a predetermined potential, is sometimes not accompanied by a variation of the polarization curve, since the current is not spent on metal dissolution under cathodic polarization. A rather similar situation may also occur under anodic polarization, e.g., when the corrosion rate rises simultaneously with oxygen evolution on the metal surface, as occurs on carbon steel in alkaline solutions (see No. 1 in Table 8.2).

The modulus of the potential shift to the negative side under the action of a cathodic current is usually smaller than the modulus of the shift to the positive side, by an anodic current of the same size. However, this difference can be regarded in the given case as quantitative, but not as qualitative. Therefore, the above reasoning is applicable also when corrosion attack by external cathodic currents is considered.

REFERENCES

1. N. T. Vagramian and T. V. Il'ina-Kakueva, Current Distribution over the Surface of Electrodes during Electrodepositing of Metals (in Russian), Metallurgiya, Moscow, 1955, 68p.
2. Z. A. Foroulis, Anti-Corrosion Methods and Materials, 1980, Vol. 27, No. 11, 5–8, 10, 13.
3. I. P. Poddubny and N. P. Gnusin, Theory of Polarization Fields, Collected articles "Electric Fields in Electrolytes" (in Russian), Nauka, Novosibirsk, 1967, pp. 3–14.
4. Yu.Ya. Iossel and G. E. Klionov, Zashchita Metallov, Vol. 9, No. 6, 1973, 635–649.
5. R. H. Rousselot, Repartition du Potentiel et du Courant dans les Electrodes (in French), Dunod, Paris, 1959, 80p.
6. A. Travulsko, J. Electrochem. Soc. 1974, Vol. 121, No. 5, 660–665.
7. V. P. Mashovetz and V. G. Fomichov, Jurn. Fiz. Khimii, Vol. 34, No. 8, 1960, 1795–1801.
8. Yu. Ya. Iossel, G. E. Klionov and R. A. Pavlovsky, Computation and Simulation of Contact Corrosion of Ship Structures (in Russian), Sudostroenie, Leningrad, 1979, 261p.
9. I. V. Riskin, Ya.B. Skuratnik, L. M. Lukatsky et al., Zashchita Metallov, Vol. 19, No. 6, 1983, 899–908.
10. I. V. Riskin and Ya. B. Skuratnik, Zashchita Metallov, Vol. 20, No. 1, 1984, 97–102.
11. V. S. Daniel-Bek, Jurn. Fiz. Khimii, Vol. 22, No. 6, 1948, 697–910.
12. A. N. Frumkin, Jurn. Fiz. Khimii, Vol. 23, No. 12, 1949, 1477–1482.
13. O. S. Ksenjek and V. V. Stender, Dokladi Akademii Nauk SSSR, Vol. 107, No. 2, 1956, 280–283.
14. O. S. Ksenjek and V. V. Stender, Jurn. Fiz. Khimii, Vol. 31, No. 1, 1957, 118–127.
15. J. Newman and C. Tobias, J. Electrochem. Soc., Vol. 109, No. 12, 1962, 1183–1191.
16. I. V. Riskin and Ya. B. Skuratnik, Zashchita Metallov, Vol. 21, No. 2, 1985, 236–243.

CHAPTER 9

ELECTROCORROSION PROTECTION OF METALS IN ELECTROCHEMICAL PLANTS BASED ON EXISTING METHODS

Contents

The methods and means of protection of metallic piping and structures from electrocorrosion in electrochemical plants considered in this chapter are similar, at least partly, to the ones used for the protection of underground metallic structures from corrosion by stray currents. Modifications of these methods which take into account the distinguishing features of electrochemical plants and sometimes completely transform the existing methods are also discussed.

9.1. INSULATING COATINGS

Since it is impossible to stop the electrolyte flows inside the operating technological lines, the leakage currents that penetrate to these lines from the electrolyzers were always regarded as an unavoidable harm. Therefore, up to the 1980s, the protection of the metallic piping and equipment in electrochemical plants from electrocorrosion mainly consisted of their isolation from contact with the aggressive environment and of other means of reducing the leakage current. As mentioned in

Chapter 2, similar means are applied for the protection of underground and underwater structures from corrosion by stray currents.

Isolation of the metal from contact with the aggressive environment with the help of non-metallic, corrosion-resistant materials is also used in many cases for the protection of metallic structures from attack by leakage currents in electrochemical plants, although most of the existing paints and polymer coatings are penetrable by water and oxygen [1]. It is obvious that the stability of a structure protected by an isolating coating is determined by the stability and quality of this coating, and not by the characteristics of the isolated metal. Therefore, carbon steel is most often used as the coated base metal.

Underground and underwater metallic structures may continue working for quite a long period of time, even in the case of partial damage to the coating, if they do not come under the attack of stray currents. Cathodic protection of the structures that have defects in their coatings further increases their lifetime.

In relatively slow corrosion processes, such as the corrosion of carbon steel in neutral media, defects and pores in the coating may become clogged by corrosion products, which leads to a dying down of the corrosion processes in the defects. On the other hand, in media of high aggressiveness, at elevated temperatures and flow rates, the corrosion products usually dissolve or are removed by the solution flow, and the corrosion rate does not go down with time.

A combination of high environmental aggressiveness, severe technological regimes and the presence of external currents determines the higher requirements demanded from the isolating materials and the isolation quality of the metallic piping and equipment in the electrochemical plants, which limit the range of appropriate materials for selection.

As the external current concentrates in the defects of the coating, the current density in these defects may become very high. Accumulation of the corrosion products leads to bloating end, consequently, to mechanical destruction of the coating in the defective areas. Therefore, along with the requirement of a high chemical resistance, the most stringent demands of tautness, strength and non-porosity are imposed on the protective coatings in electrochemical plants.

Enamels are unsuitable for the protection of piping and equipment in the zones of external currents because of their porosity. An attempt to apply an enamel coating for the protection of a 1 km long and 100 mm diameter brine supply pipeline of carbon steel in a diaphragm electrolysis plant failed: the whole line was replaced by a Teflon pipeline after 4 days of operation. The major part of the enameled pipeline failed already on the first day. In the absence of leakage currents, other things being equal, this line should be in operation for a long time.

Rubber linings and some other kinds of lining belong to non-porous coatings that are able to work without destruction for a relatively long time [2]. For example, a fairly positive effect was experienced using chlorine-resistant linings, such as ebonite, in chlor-alkali plants [3]. Ebonite produced on the basis of chloroprene rubber is the most stable in the environments of these plants [4]. Rubber lining has been used for many years for the protection of pipelines, electrolyzer covers and other equipment [5]. Large-size equipment was protected by linings with an underlayer of ebonite or of another polymer material and 2–3 layers of diabase tiles [3].

Practical experience has shown that the lifetime of the isolating coatings depends not only on the environmental aggressiveness, but also on the quality of the isolating materials, and it can thus vary in a large range, from 0.5 to 5 years [3] under the same working conditions.

It was noted in previous chapters that oxidizers belong to the most common environmental components of the electrochemical plants. Their presence leads to a sharp reduction in the lifetime of polymeric materials and of rubber linings.

The lifetime of rubber-lined piping for wet chlorine and chloranolyte does not usually surpass 1 year, and for the brine piping (which does not contain dissolved chlorine) it is significantly higher: 1.5–3 years. The lifetime of the rubber-lined covers of electrolyzers is also fairly limited due to destruction of the lining, which preferentially occurs at the boundary between the electrolyte and wet chlorine phases [5].

Although the problem of selecting accessible, reliable and stable isolating materials for the highly aggressive environment of electrochemical plants has not found a conclusive solution until now, using these protective materials still remains attractive. This is explained by the comparatively low cost of this protection and by the possibility of the anti-corrosion crews of the electrochemical plants executing such protection themselves.

9.2. REDUCTION OF LEAKAGE CURRENTS COMING FROM ELECTROCHEMICAL CELLS

Reduction of the external current acting on the metallic structure is usually regarded as a means of diminishing, but not of preventing, the corrosion destruction of the metal. In fact, when a metal corrodes in an active state (as, e.g., carbon steel in soil or water) the corrosion rate is proportional to the magnitude of the external anodic current, i.e., the lower the current, the lower the corrosion rate. Moreover, reducing the external current acting on the passive metals may, in some cases, completely prevent the corrosion failure. Such a possibility is proved in the example on titanium in a chloride–alkaline solution considered in Chapter 8 (see Table 8.1, No. 3). When the external current I_e is reduced to below 1.4 A, the criterion $I_e/I_{lim} < 1$ is met, and the titanium structure remains stable. This became possible due to the relatively high value of the current density in the passive state of titanium in this solution. In any event, even when a reduction of the external current does not lead to the satisfaction of condition $I_e/I_{lim} < 1$, a reduction in the corrosion rate is achieved. Therefore, the attempt to reduce the external current is always justified.

Reduction of leakage currents which penetrate from the electrolyzers to the pipeline and equipment is usually achieved by increasing the electrical resistance between the electrolyzers and the pipelines and of the resistance to current flow along the group and set headers of the electrochemical plants [6, 7].

There are some differences between the means of reducing leakage currents along the piping for electrolytes and those for wet gases.

9.2.1. Reduction of leakage currents along electrolyte piping and prevention of current oscillations

The most simple and accessible and, therefore, the most widely used means of reducing leakage currents from electrolyzers consists of increasing the length of the tubes connecting the electrolyzers and the group headers. When these tubes are made of insulating materials, the electric resistance of the electrolyte flowing inside them is proportional to their length. These means are most feasible when the tubes of insulating material, such as rubber, have a comparatively small diameter (25–50 mm). For example, a quite effective reduction of the leakage current along the brine pipelines of a diaphragm electrolysis plant was achieved by increasing the length of the rubber hoses used for feeding the brine to the electrolyzers [8].

Feeding nickel catholyte to the electrolyzers in nickel electrorefining plants is carried out through distribution rakes [9]. Long (more than 1 m) rubber hoses are attached to the rake nipples which have a diameter of 6–10 mm.

Increasing the path length of the electrolyte at the degassing tanks output (from which the electrolyte is sent to the electrolysis baths) in copper electrorefining plants can be achieved by improving the construction of the distribution rakes (Figure 9.1). The path length in the improved device is increased by 1–4 m; in addition, the electrolyte section inside the device is reduced [10].

At the areas of the electrolyte outflow, it is possible to raise the electric resistance by breaking the jets of the electrolyte into drops or by dividing them into smaller jets. A lot of devices have been developed for this purpose, including the ones installed at the input and output areas of the electrolyzers [11–16]. Theoretically some of these devices could produce an effective breaking of the jet. However, in practice they are not always sufficiently effective, for different causes: fouling and clogging by salt depositions, skewness of parts of the device, etc. Nevertheless, some of the devices for jet breaking are widely applied in practice. For example, devices for jet breaking are applied on the caustic liquor lines at the output of diaphragm electrolyzers [5]. When care is taken in the maintenance and control of these devices, they enable a reduction of the leakage currents from the electrolyzers.

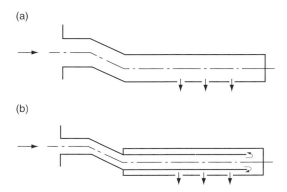

Figure 9.1 Distribution rake of insulating material: (a) old construction; (b) improved construction. Arrows show the direction of the electrolyte flow.

However, their efficiency diminishes as the capacity of the electrolyzers rises, since it is more difficult to break the jets of an larger diameter pipe.

As already noted above, the values of the leakage currents and their distribution character largely depend on the layout of the piping and technological equipment [6, 7]. The location of the electric resistance along the technological line also influences the magnitude of the leakage currents. For example, in work [6] it is shown that in some technological lines, increasing the resistance between the pressure tank and the header reduces the leakage currents more significantly than increasing the resistance between the connecting tubes of the electrolyzers. Therefore, between the metallic group and set headers, as well as in some areas of the headers themselves, it is expedient to install insulating inserts. Replacement of these inserts in the case of their destruction as a result of their low chemical resistance, is much less expensive than replacing the whole piping. Sometimes, increasing the electric resistance along the piping is executed by breaking metallic contacts between the connecting flanges. As in the cases of prevention of contact corrosion, contact breaking is carried out by insulating the connecting bolts from the flanges with sleeves and washers of insulating materials that are placed on the bolts [17].

Intensive corrosion was revealed inside the gas ducts for the flow of the gas–liquid mixture in the electrolyzers of water electrolysis plants [18]. The corrosion was caused by current oscillations inside the electrolyte which was saturated with gas bubbles. Prevention of this corrosion damage was achieved by the installation of expanders [19] at the connection tubes of the electrolyzers. These devices homogenized the gas flow, suppressed the current oscillation and, in this way, prevented the development of corrosion.

It is shown in Ref. [20] that modulation of current oscillations in alkaline water electrolysis noticeably influences the activity and stability of nickel cathodes. It may be expected that excluding this oscillations would increase the efficiency and operation stability of the cathodes.

9.2.2. Reduction of leakage currents along piping of wet gases

As the sources of the leakage currents along the wet chlorine piping in chlor–alkali plants were considered, it was shown that the condensate layer which is formed on the internal surface of the tubes (chlorine taps) made of an insulating polymeric material connecting the electrolyzers and the headers, constitutes a pathway for the leakage currents coming from the electrolyzers to the pipelines and equipment. The condensate of the wet chlorine contains about 1 g/l of NaCl and dissolved chlorine. The internal surface of the chlorine taps is chlorinated and becomes porous. Due to the presence in the condensate layer of the technological solution components and as a result of impregnation of the internal layers of the chlorine taps with this condensate, these tubes have a fairly high conductivity.

In water electrolysis plants, the condensate is alkaline. Gaskets of rubberized asbestos which are installed at the flange connections of the hydrogen piping are impregnated with alkaline condensate and become conductive [21]. All these circumstances lead to the penetration of quite high leakage currents to the piping and equipment of the gas lines in water electrolysis plants.

Although the leakage currents along the wet gas lines are lower than those along the electrolyte lines, they can cause intensive corrosion of metallic piping and equipment, as occurred with the titanium headers of the wet chlorine lines in chlor–alkali electrolysis plants (see Chapter 4). Therefore, the wet gas lines, like the electrolyte lines, must be supplied with a means of reducing the leakage currents.

As for the electrolyte lines, increasing the length of the connecting tubes is also applicable for the wet gas lines. For example, in diaphragm electrolysis plants the length of the chlorine taps attains 2–3 m. Insulating inserts are also often installed in the group and set headers of these plants.

Some special means for reduction of the leakage currents can be used in the wet gas lines. First of all, the correct selection of the material and structure of the tubes connecting the electrolyzers with the headers must be done. Along with high insulating properties, the material of the connecting tubes must possess a high chemical stability and low wettability. The best material possessing these properties is polytetrafluoroethylene (Teflon). This material does not become friable and porous on contact with the aggressive gas. When the flowing layer of condensate contacts with Teflon, which is characterized by its low wettability, discontinuities are formed in the flow. This leads to an increase in the ohmic resistance of the condensate layer.

Accumulation of condensate inside the connecting tubes must be avoided. For this purpose it is recommended that the horizontal sections of the connection tubes be provided with a slight slope and that deflections and shoulders on the internal tube surface be avoided.

It is seen from Table 9.1 that a sharp reduction of the leakage currents was achieved when Teflon tubes replaced the rubber lined tubes of the chlorine taps. In most of the Teflon chlorine taps the leakage currents were reduced by two orders of magnitude or more. The electrical resistance of the Teflon chlorine taps determined the character of the distribution of the leakage current along the set of electrolyzers. The typical character of current distribution (going down from the margins to the center of a set) was not observed in this case. As seen from Table 9.1, even the leakage currents from adjacent electrolyzers may have opposite directions.

Testing titanium headers with Teflon chlorine taps have shown that the first corrosion perforations on titanium appeared after no less than 1.5 years of operation. It is most probable that the damage resulted from a growth in the current, only

Table 9.1 Leakage current along wet chlorine taps made of rubber lined carbon steel and of Teflon

No. of electrolyzers	Leakage current (A)	
	Rubber lining	Teflon tube
1	4.8×10^{-3}	3.0×10^{-5}
3	1.05×10^{-2}	2.3×10^{-5}
13	4.5×10^{-4}	7.1×10^{-5}
22	1.5×10^{-2}	-4.5×10^{-6}
126	-7.5×10^{-3}	3.0×10^{-4}
128	-7.5×10^{-3}	-1.5×10^{-7}

at the end of this period of time. Deposits which formed on the internal surface of the tubes increased the wettability of the Teflon surface, which led to the current growth. It may be suggested that regular cleaning or timely replacement of the Teflon tubes should increase still further the operation time of the titanium headers without any corrosion damage. Thus, the correct selection of the material and the structure of the connecting tubes for wet gases can be considered as an effective means of metal protection against corrosion attack by leakage currents. However, it is impossible to completely prevent the corrosion damage by these means.

The application of insulating inserts made of materials such as Teflon in horizontally disposed headers for wet gases is less effective than those in vertical and sloping sections of the headers, because accumulation of condensate is possible between the shoulders of the flange connections of horizontal pipes. A special device was developed for interrupting the condensate layer in horizontal sections of metallic pipes for wet gases. The efficiency of this device was studied in a laboratory installation which modeled the operating conditions of a pipe for the transportation of wet chlorine (Figure 9.2).

Chlorine gas was heated and wetted in vessel 1 containing distilled water, and in vessel 2a containing 100 g/l NaCl solution and introduced into rubber tube 3 through lateral orifices made in its wall and disposed above the solution. The chlorine then passed through a short titanium tube 4, connecting rubber tube 5, titanium tube 6, second connecting rubber tube 7, second titanium short tube 8, and rubber tube 9. The lower ends of tubes 3 and 9 were immersed in the 100 g/l NaCl solution of vessels 2a and 2c. From the output of vessel 2c, the chlorine was sent for absorption by an alkaline solution.

The lengths of titanium tubes 4 and 8 were 60 mm, the length of titanium tube 6 was 1000 mm; the internal diameters of all the regarded tubes were 10 mm.

Vessels 1 and 2a were thermostatted at a temperature of 90°C. Graphite electrodes 11a and 11c were immersed into the solutions of vessels 2a and 2c, respectively. During the flowing through the system of tubes 3–9, heated and

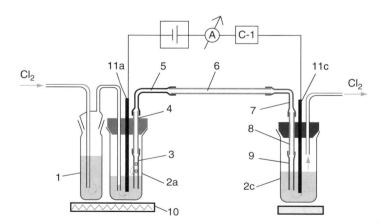

Figure 9.2 Installation for modeling operation conditions of titanium piping for wet chlorine in the field of external current (explanations are given in the text) (see color plate 2).

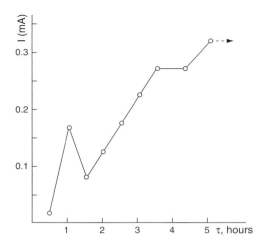

Figure 9.3 Character of current growth along the titanium tubes simulating titanium piping for wet chlorine.

wetted chlorine formed a layer of condensate on their walls, which produced an electric circuit between the electrodes 11a and 11c. The current of this circuit was registered by a milliammeter. A coulometer C-1 measured the current quantity that passed through the circuit during the experiments.

The initial current appeared 20–30 min after supplying a voltage of 350 V across the electrodes. During 5 h of testing the current reached a magnitude of 0.3 mA, and after that it oscillated around this value (Figure 9.3). It seems that the formed condensate layer attained a maximum thickness over 5 h. Current that was close to the maximum value (0.3 mA) reappeared almost immediately upon a restart (after the previous 6–8 h of work). This means that the condensate layer formed inside the chlorinated rubber tubes, remained inside them after the previous tests.

After 25 days of testing at the considered installation, pits filled with white corrosion products (TiO_2) were formed at the ends of titanium tubes 4, 6 and 8 disposed at the side of the cathode, i.e., attacked by the anodic current (Figure 9.4). A comparison of the weight loss (0.011 g) with the magnitude of the electric current passed through the circuit (160 Coulombs) showed that no less than 60% of the current was spent on metal destruction. The total damaged surface area attained was 0.1–0.2 cm^2. Hence, at a current of 0.1 mA (about 1 h after the beginning of the test) the metal was in an active state (pitting), since the current density, assessed in relation to the damaged surface area, reached the value of about 1 mA/cm^2. As was shown in Chapter 6, the titanium activation potential E_a was attained under similar conditions at much lower values of the current density. After a restart, the conditions of titanium activation were reached almost immediately. Thus, the operating conditions of the titanium piping in the field of the external currents can be simulated in the considered installation.

Studies of the device for interrupting the condensate layer were carried out in the considered installation. The operating principles of the device (Figure 9.5) involve

(a)   (b)

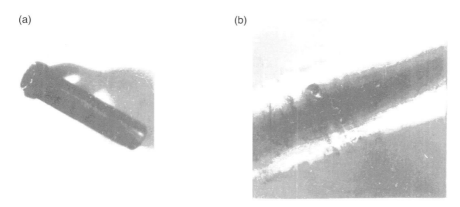

Figure 9.4 Corrosion damage of sections of the titanium tubes attacked by anodic leakage currents: (a) – at the tube end; (b) – under the rubber hose.

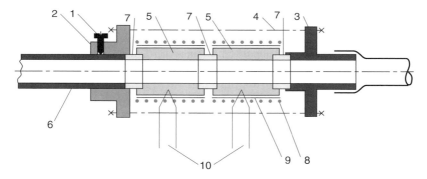

Figure 9.5 Device for condensate layer interruption (explanations are given in the text). (see color plate 3).

heating a pipe section, which is insulated from the rest of the pipe, to a temperature at which the condensate evaporates from the heated surface area [22].

The tested model of the device included two titanium bushes 5 insulated from one another by a Teflon ring 7. Similar rings 7 were installed at the other ends of the bushes 5. Gas-tight connection of the bushes 5 and rings 7 was executed by adjusting flanges 2 and 3, which were tied by studs 4. Spirals 8 of Ni–Cr alloy were put on bushes 5 and insulated from the bushes by mica pads 9. The wall temperature of the bushes 5 was controlled by thermocouples 10.

The device was mounted at one of the ends of tube 6 of the considered installation (see Figure 9.2) with the help of a screw 1. In the process of stepwise temperature increments (by 10°C), the magnitudes of the currents between the electrodes were measured.

The results of the measurements after heating to a temperature of 150°C are presented in Figure 9.6. During the initial heating period, some current growth

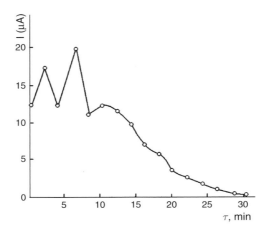

Figure 9.6 Current reduction along titanium tubes of a model device for condensate layer interruption in chlorine piping.

occurred, owing to the increasing conductivity of the condensate layer with rising temperature. Then, over a short period, oscillations in the current were observed, related, most probably, to the unstable state of the condensate layer at the initial stage of its evaporation. After 5–10 min, when a temperature of 150°C was reached at the internal surface of the titanium bushes, the current began to diminish and dropped to a minimum magnitude of 0.5 μA over 30 min, which was about 30 times lower than the initial magnitude of the current.

As the heating of the bushes was stopped, a reduction in the current from 0.5 to 0.1 μA took place, which can be explained by a reduction in the ion conductivity of the dry deposit formed on the internal surface of the Teflon rings with the reduction in temperature. Twenty minutes after the heating was stopped, the current began to rise anew, and reached its initial magnitude.

The titanium bushes did not corrode either at 150 or at 170°C. These results corroborated the data [23] on the corrosion stability of titanium in wet chlorine up to a temperature of 200°C, i.e., in the presence of a superheated vapor.

There are data [24] indicating that the moisture content sufficient for providing titanium stability in chlorine at temperatures up to 135°C is 0.005%; according to more recent information [23] this required moisture content is 0.98%. At a temperature of 95°C, 1 kg of chlorine that comes from the diaphragm electrolyzers into the chlorine pipeline contains about 1.3 kg of water [25]. Hence, at temperatures of up to 150°C, titanium will retain its corrosion stability in the wet chlorine lines.

Thus, interruption of the condensate layer by heating the insulated sections of the piping to a temperature of about 150°C can be applied as an effective means of protection from corrosion by the attack of external currents for metallic piping for wet gases in electrochemical plants.

When such devices are applied in practice, measures should be implemented to avoid penetration of the condensate to the heated sections from other areas of the piping. The pitch of the pipes has to be designed in such a way as to promote the flow of the condensate along the piping. The device would be much more reliable and effective if a ring shoulder is provided at the input of the insulated section. In the lower part of the pipe, before this shoulder, a condensate drainage system must be installed.

Devices for interruption of the condensate layer must be provided with a means for monitoring and controlling the internal surface temperature of the heated sections, including the insulating rings. As the device works under the conditions in which sparking is forbidden, heating can be carried out with the help of super-heated vapor.

Since the devices for interrupting the condensate layer are rather complicated and require constant control, they can be installed in only a limited number of pipeline sections. It is advisable to combine the use of these devices with other means of protection of the piping from electrocorrosion. For example, when a protection similar to electrodrainage is executed (connecting the header to be protected to the negative pole of a current source or rectifier), the device has to be installed in the marginal section (output) of the set header. In this area, the device interrupts the path of the current flow through the condensate layer from the header to the metallic equipment located outside the electrolyzer set, and thus prevents the electrocorrosion damage of that equipment. Therewith, lateral paths of current penetration must also be prevented. The possibility of this type of protection will be considered below.

9.3. SECTIONALIZATION OF PIPING MADE OF PASSIVE METALS

In Chapter 8 it was shown that the maximal potential value of a metal at the end of a tube-shaped structural element in the field of an external current depends on its length. Expressions were presented for determining the permittable length of monopolar and bipolar tube sections. It was shown, in particular, that it is possible to prevent the potential shift of a metal to the activation potential value E_a at the anodic end of a bipolar tube, i.e., reaching the fulfillment of condition 8.1, by reducing the tube length to a permittable size. This suggests the idea that sectionalization of the piping by creating a set of sections of tubes insulated from one another, can be used under certain conditions as a means of corrosion protection in the field of an external current [26].

Sectionalization of pipes made of passive metals differs in principle from the sectionalization of underground and underwater pipelines considered in Chapter 2 [27]. In the latter case, a reduction in the corrosion rate is only achieved of actively corroding carbon steel; however, the number of areas attacked by external currents increases in accordance with the number of insulated sections.

In some very simple cases, the possibility of sectionalization can be assessed without any special calculations. For example, if the voltage in a set of electrolyzers for copper electrorefining is 320 V, and the activation potential of titanium in the technological solution is 140 V, then, assuming a uniform potential distribution along the titanium technological pipeline located parallel to the set, protection can be provided by dividing the pipeline into three or more equal sections insulated from one another. Consequently, under these conditions, sectionalization of pipelines made of metal with a high E_a value and acted on by an external current can be an effective means of protection from corrosion, without requiring the application of any additional means and methods of corrosion protection.

The voltage across a bath for the chlor-alkali electrolysis of common salt is about 3.5 V [28], i.e., below the activation potential of titanium $E_a \approx 6$ V. Consequently, if a single bath has a titanium frame, no measures are needed to protect the titanium from corrosion by an external current. However, a titanium pipeline for a set of 120 baths must be divided into about 100 sections, which is inconvenient to implement in practice.

Note that in sectionalization the values of E_a must be taken into account, in particular, at the sites where the external currents act. These sites are usually the flanged joints, which have welding seams and where crevices can form under the packing. According to the data given above, the potential E_a for titanium under the influence of an anodic current in the welding seam zones and in crevices, can go down sharply. Specifically, in the electrolysis of nickel (sulfate–chloride solution at 85°C) the E_a goes down from 85 to 5.4–5.6 V both in crevices and at welding seams. In the solution for copper electrorefining (sulfuric acid solution at 65°C) the E_a at a welding seam is reduced from 140 to 9.5–10 V.

In practice, the distribution of potentials and currents along a branched system of pipes and associated metallic and non–metallic equipment is not uniform along the set of electrolyzers, and it is difficult to take account of the many particular features of each specific technological scheme in the calculations. Therefore, it is most expedient to consider the distribution of the potential in the separate pipe sections, which can be considered as a bipolar tube.

A solution to the problem of the permissible length L of a bipolar pipe is given in work [29] and in the previous chapter of this volume. This solution is significantly simplified if the activation potential value E_a surpasses the value of the stationary (open circuit) value E_s of the metal by some orders of magnitude ($E_a \gg E_s$). Only in this case, does the sectionalization of the piping become expedient. The magnitude of E_s usually varies in a narrow range and does not surpass the value of 1 V, even in environments which have a high redox potential. Hence, the value of the initial (open circuit) potential E_s can be ignored and the maximum potential shift to the anodic side of the tube section (ΔE_{max}) can be accepted, with sufficient accuracy for our purposes, as close to the E_{max} value, i.e.,

$$\Delta E_{max} = E_{max} - E_s \approx E_{max}.$$

The value of the activation potential E_a can be accepted as the maximum permissible potential, i.e., $E_{max} \leq E_a$.

Typical anodic and cathodic polarization curves for a passive metal that is characterized by a high activation potential value can be represented, respectively, by curves 1 and 2 in Figure 9.7. The current density value i_p in the anodic curve 1 in the potential range from E_1 (close to E_s) to E_a is constant, or its variation is insignificant and can be ignored.

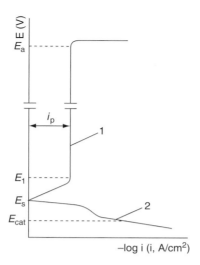

Figure 9.7 Typical anodic (1) and cathodic (2) polarization curves for a passive metal used for the calculation of piping sectionalization.

The deviations in the potential from E_s to E_{cat} at the cathodic end can also be ignored, since $|E_{cat}| << |E_a|$. The most probable values of E_{cat} lie in the range from 0 to -1.5 V, where the most negative value corresponds to the potential of water decomposition. If water contains components capable of being reduced on the metal, the maximum cathodic potential will be even more positive. The length of the cathodic end of the section can therefore be ignored in the considered case. This is all the more justified because the variation of the potential at the cathodic end is more rapid than at the anodic end, since the cathodic polarizability of a passive metal is usually lower than the anodic one (see Figure 9.7), while the rate of change of potential increases with decreasing polarizability [30].

An expression for determining the permissible length L of a tube section with a radius r in the field of an external current I_e was obtained in Ref. [26]. The parameters of the passive metal: E_s, E_1, E_a and i_p, the values of which can be found from the anodic polarization curve, are used in the expression. The final equation takes the following form:

$$L = \frac{(E_1 - E_s)\pi r^2}{\rho\sqrt{I_e^2 - \dfrac{4\pi^2 r^3 i_p (E_a - E_1)}{\rho}}} + \frac{I_e - \sqrt{I_e^2 - \dfrac{4\pi^2 r^3 i_p (E_a - E_1)}{\rho}}}{2\pi r i_p}, \qquad (9.1)$$

where ρ is the conductivity of the solution.

This expression (9.1) affords a rapid and reliable means of estimating the maximum safe length of a section of pipeline or other tube-shaped structure made of passive metal.

9.4. Possibilities of Applying "Traditional" Methods of Electrochemical Protection and Their Modifications for Electrochemical Plants

As the condition (8.1) of the corrosion stability of a structure consists of maintaining the metal potential within the predetermined permissible boundaries, it can be suggested that the necessary stability can be achieved by using electrochemical methods of corrosion protection.

Cathodic protection has a limited application in environments of high aggressiveness owing to the elevated energy expenditure (protection by an impressed current) and to the intensive consumption of the sacrificial anodes [31]. It is even more ineffective under conditions of attack by external currents, where these expenses further increase. Anodic protection, which is quite widely used in chemical and other branches of industry [32] needs very strict control of the potential and current values, which cannot be done in the presence of uncontrolled external currents.

9.4.1. Electrodrainage protection

Electrodrainage, which is applied as a major method of corrosion protection against attack by stray currents, can also be used for the protection of metallic structures in electrochemical plants. For example, the frames of electrolyzers for electrolytic borating were protected by electrodrainage [33]. In this case, the frames were connected to the negative pole of the power source.

A method for the corrosion protection of some metallic structural elements connected with electrolyzers was proposed in a Polish patent [34]. It can be considered as electrodrainage since it consists of using the positive pole of a power source (rectifier), to which is connected an auxiliary electrode (anode), through a resistor. In an example which is given in this patent, sections of tubes made of acid-resistant steel for the input and output of electrolyte in a copper electrorefining plant were protected. A short tube of lead was used in this example as the auxiliary electrode. As the protection was executed, the potential of the acid-resistant steel shifted to the negative side, from 1.1 to 0.2 V, i.e., from the transpassive field (see Chapter 1) to the field of a stable passive state. In this example, as in other cases of electrodrainage protection, the size of the current coming from the electrolyzers to the communication lines sharply increases, owing to a fractional shunting of the current path through the electrolyte by metallic conductors. This type of protection can be applied in some special cases which are defined by specific characteristics of the technological design and by structural parameters of the protected equipment.

A similar protection principle to the one considered above was proposed in a Czechoslovak patent [35] for the protection of pipelines and equipment in the areas in which corrosion damage is possible by leakage currents coming from anodic protection installations. In this case, the auxiliary electrode was connected to the positive pole of the power source used for anodic protection. The auxiliary electrode, insulated from the communication lines was, installed in the protected area.

The possibility of using electrodrainage protection for titanium headers was tested in a chlor–alkali diaphragm electrolysis plant during a period when only two of the ten groups of electrolyzers were in operation. The insulating inserts of rubber-lined carbon steel installed between each of the group headers and the set header were shunted before testing, to prevent electrocorrosion damage to the areas of the set header that were adjacent to these inserts. The set header was directly connected to the negative pole of the series power source (rectifier) by a metallic conductor, and the measured amperage along this conductor was equal to 22.4 A. The general load of the commercial electrolysis plant in the period of the measurements was 40 kA. Thus, in the absence of any additional resistors in this circuit, the magnitude of the drainage current came to only 0.05–0.06% of the electrolysis current. This result suggests that, in principle, electrodrainage can be applied to protect titanium headers of wet chlorine. However, it must be taken into account that this protection may be used only in the absence of any connection of the electrically insulated protected structure with other structures through the electrolyte. Moreover, the metal of the protected structure has to be resistant against hydrogenation and corrosion at the negative potential values to which it shifts under the conditions of the electrodrainage protection.

In many cases, it is difficult, and sometimes even impossible, to prevent the connection of the protected structure with adjacent unprotected metallic structures through the electrolyte. For example, corrosion damage of heat exchangers, valves and other metallic equipment located out of the electrolysis zone may occur when electrodrainage protection is used. It is practically impossible to break the electrolyte flow inside the systems that are filled with the electrolyte. In addition, the leakage currents in these systems can grow to values at which energy losses become significant.

In wet gas lines, the interruption of the electrolyte layer can be executed with the help of the device that was described in Section 9.2.2 of this chapter.

It can be concluded that the traditional methods of electrochemical protection and their modifications, in which external power sources are used, can find only a limited application in the protection against electrocorrosion in electrochemical plants.

9.4.2. Protection by sacrificial anodes for current drainage in chloride electrolysis plants

Sacrificial anodes made of actively dissolving metals for current drainage, which are applied in sectionalized piping as one of the means of the protection against corrosion attack by stray currents of underground structures of carbon steel and of the reinforcement in concrete, were discussed in Chapter 2. The stray current runs off from the insulated section to these anodes, and from the anodes it drains off into the ground. Thereby, the anodes are dissolved by the current and prevent corrosion damage to the structure.

It is clear that such anodes are inapplicable for the protection of piping and equipment in the highly aggressive environments in electrochemical plants. This is due, not only to the difficulties of installing them inside the operating equipment and to the necessity of executing constant control and monitoring of these anodes, but also to contamination of the technological solutions by corrosion products of these anodes.

Protection by sacrificial titanium anodes for current drainage, which was used in one of the electrochemical plants can be considered as a modification of this means of protection. These anodes were installed at the branches of a titanium group header for wet chlorine attached to the chlorine taps made of insulating material connecting the header with the electrolyzers (Figure 9.8).

Replaceable anode 1 was mounted between branch 2 of the header and chlorine tap 3, made of insulating material, with the help of adjustable flanges 4 and studs 5. Sealing of the connected parts was provided by rubber gaskets 6. The electric connection between the anode and the branch was carried out through titanium contacts 7, connected by a titanium bolt 8. The area damaged by the external current was displaced from the fixed branch 2, to an easily replaceable anode 1. As the chlorine gas constitutes the technological environment in this case, the problem of environmental contamination did not arise.

However, the lifetime of these anodes was short; from several weeks to several months (Figure 9.9). Regular operations of control and replacement of the destroyed anodes was necessary, which interrupted the normal operation of the equipment. Moreover, this form of protection led to significant losses of expensive metals, such as titanium, and required additional operations for manufacturing new anodes. Therefore, this protection can be regarded as a temporary means, to be carried out only in cases when other, more effective, methods of protection cannot be applied. However, it can be regarded as much more effective when such sacrificial anodes are used in combination with Teflon chlorine taps.

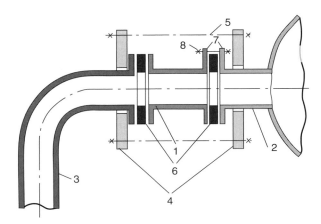

Figure 9.8 Protection of a branch of titanium group header for wet chlorine by titanium sacrificial anode for leakage current drainage (explanations are given in the text) (see color plate 4).

Figure 9.9   Perforation of titanium sacrificial anode for leakage current drainage.

9.4.3.   Protection by sacrificial anodes for current drainage in electrolysis plants with metal deposition

Protection against corrosion attack by anodic leakage currents with the help of sacrificial anodes for current drainage, can also be applied in electrolysis plants with metal deposition, such as electrorefining plants. In this case, the principle of the protection involves the installation of anodes made of electrorefining metal in the areas of anodic current action. As the protected area of the metallic structure and the anode are brought into electrical contact, the metal dissolution under the action of the external anodic current is displaced from the protected metal to the actively dissolving anode. The current that drains off from the anode is spent on its dissolution. This protection is not accompanied by electrolyte contamination because the anode is made of the same metal that is dissolved in the industrial electrorefining process.

Since the technological solutions for electrorefining are so selected that the refining metal dissolves at a minimal overvoltage, the major share of the external anodic current acting on the structure of the passive metal is concentrated on the anodes.

Let us consider the possibility of applying this principle for the protection of titanium and stainless steel 18-10 under the conditions of copper electrorefining and for the protection of titanium under conditions of nickel electrorefining. As seen from Figure 9.10, at the potentials of the active dissolution of copper and nickel (curves 1–3), titanium and stainless steel are in a stable passive state (curves 4 and 5). The current densities at the passive state of these latter metals are 3–4 orders of magnitude lower than the current densities of the active dissolution of copper

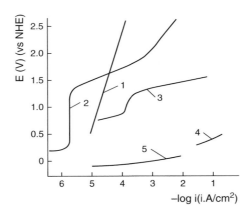

Figure 9.10   Anodic polarization plots on titanium (1, 2), stainless steel 18-10 (3), copper (4) and nickel (5) in electrolytes of electrorefining of copper (1, 3, 4) and nickel (2, 5).

and nickel. This indicates the possibility of applying this principle of protection in the conditions under consideration.

Galvanostatic tests using the combined electrodes, steel 18-10–copper and titanium–copper, in the electrolyte of copper electrorefining, and titanium–nickel in the electrolyte of nickel electrorefining, were carried out to verify the efficiency of this protection principle. Each of these electrodes consisted of two 10-mm diameter cylindrical specimens made, respectively, of protected and of refining metals. For connecting this pair of specimens, a blind center-screw hole was made at the end of one, and a center-screw tail at the end of the other. The tail was screwed into the hole. A ring gasket isolated the connected end surfaces of the specimens from penetration by the solution.

The tests were carried out at a current density of 5 mA/cm² (related to the surface area of the anode) and at a ratio of the working surface areas of the protected metal to the surface area of the anode of 2.5:1. For both titanium and stainless steel, the open circuit potential that was established before the polarization was close to the potential of the anode. At the contact with copper and nickel, this was equal, respectively, to 0.37 and 0.095 V. The protected metals retained their metallic shine and their weight losses were close to zero. The current was entirely spent on the dissolution of copper and nickel, with an efficiency close to 100% (taking into account the dissolution in the form of Cu^{2+} and Ni^{2+} ions). The corrosion of these metals was uniform and its rate reached values, respectively, of 66 and 55 g/m²h. Thus, the tests confirmed the efficiency of the considered method of corrosion protection.

An important advantage of this method is its ability to protect passive metals, no matter what their values of activation potential.

It should be noted that the consumption of the anodes does not lead to any additional expenses and losses of the anode material, because the anodes for the corrosion protection, like the anodes for the electrorefining, are made of metals that are intended for dissolution. On the contrary, these anodes allow, in some cases, "the trapping" of the leakage currents and diversion of them to the metal refining process.

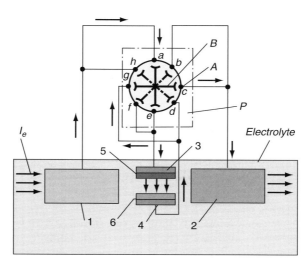

Figure 9.11 Simplified scheme of a device with two alternative electrodes for protection of metallic structures from corrosion by external currents under the conditions of electrolysis with metal deposition on the cathode (explanations are given in the text) (see color plate 5).

The dissolution rates of the refining metals in the technological solutions of the electrorefining plants are quite high, even in the absence of an external anodic current. This rate strongly increases under the influence of an anodic current, so anodes of a large mass have to be used for protection. Nevertheless, they need constant monitoring and control and must often be replaced. Therefore, despite the high effectiveness of this method of protection, the possibilities of corrosion protection by sacrificial anodes are rather limited.

Under the conditions of electrolysis with metal deposition on the cathode, the possibilities of protection by a dissolving anode can be broadened by using a special device [36], a simplified scheme of which is given in Figure 9.11.

Electrodes 3 and 4, made of corrosion-resistant conductive materials, are installed between two parts, 1 and 2, of a metallic structure, which are under the action of an external current I_e. Each of the electrodes is coated with layers, 5 and 6, of the metal which is deposited on the cathode in the given electrolysis process. Switch P of any kind (electronic, mechanical, etc.) presented in the example of Figure 9.11 is connected to electrodes 3 and 4. It consists of a stator A that has eight immobile contacts a, b, c, d, e, f, g and h and a turning cross–shaped rotor B with four mobile contacts.

The device operates in the following way:

At the position of the rotor that is indicated in Figure 9.11 by solid lines, contacts a–e and c–g are closed; thereby, electrode 3 comes into contact with part 1, and electrode 4 with part 2, of the structure. The major share of the external current I_e that leaks from the electrolyte in part 1 is directed through contacts a–e to electrode 3, and it is spent for the dissolution of layer 5 and for additional metal deposition on layer 6 of electrode 4. These processes take place because of the low overvoltage values of the anode metal dissolution and of its deposition on the same metal. In other words, at this position of the rotor, layer 5 works as a dissolving anode and layer 6 works as a cathode of the same metal. The arrows indicate the directions of the current I_e flow from the moment of its leaking into part 1 and up to its leaking out from part 2 (to which the current flows through contacts g–c) into the electrolyte.

As layer 5 is dissolved to some predetermined thickness, rotor B turns automatically by 45°, to the position indicated in Figure 9.11 by the dotted lines. Thereby, contacts a–e and c–g are interrupted and contacts h–d and f–b are closed. It is seen from the figure that, in this case, the current comes from part 1 of the structure, through contacts h–d, to layer 6, which works as a dissolving anode. At the same time, the metal is deposited on layer 5, which works as a cathode from which the current flows, through contacts f–b, to part 2 of the structure. At the following turn by 45°, the electrodes reverse their functions anew. Thus, layers 5 and 6 are deposited on electrodes 3 and 4, which alternately rise and fall in thickness.

The device can be installed inside an insulating insert disposed between the protected metallic pipes. The thickness of layers 5 and 6 is selected as a function of the size of the external current and of the surface area of electrodes 3 and 4. The device possesses all the advantages of the principle of corrosion protection by dissolving anodes, and at the same time does not require constant control and replacement of dissolving anodes. Also, it does not need the application of anodes of a large mass. Moreover, the device prevents metal deposition on the areas of the structure where cathodic currents are acting. Such deposition usually occurs in the form of dendrites, which hinder the electrolyte flow inside the pipes.

REFERENCES

1. H. H. Uhlig and R. W. Revie, Corrosion and Corrosion Control, John Wiley & Sons, New York, 1985, 464p.
2. I. Ya. Klinov, P. G. Udima, A. V. Molokanov and A. V. Goriainova, Corrosion Resistant Chemical Equipment, Handbook (in Russian), Mashinostroenie, Moscow, 1970, 591p.
3. A. M. Sukhotin and A. L. Labutin eds, Corrosion and Protection of Chemical Equipment, Handbook (in Russian), Vol. 6, Khimiya, Leningrad, 1972, 373p.
4. N. A. Mokrova, I. F. Rozen, N. S. Zaytzeva et al., Proceedings of Conference "Corrosion Protection in Chemical Industry" (in Russian), Cherkassi, 1985, p. 77.
5. V. M. Zimin, G. M. Kamarian and A. F. Mazanko, Chlorine Producing Electrolyzers (in Russian), Khimia, Moscow, 1984, 302p.
6. N. N. Shvetzov, Tzvetnie Metalli, No. 8, 1962, 36–44.
7. N. N. Shvetzov, Khimicheskaya Promishlennost, No. 1, 1965, 60–64.
8. L. M. Yakimenko, Electrolyzers with Hard Cathode (in Russian), Khimiya, Moscow, 1966, 300p.
9. Yu. V. Baymakov and A. I. Jurin, Electrolysis in Hydrometallurgy (in Russian), Metallurgiya, Moscow, 1977, 336p.
10. I. V. Riskin, L. M. Lukatsky, L. V. Zaytzeva and A. A. Kondakov, Tzvetnaya Metallurgiya, No. 2, 1983, 31–34.
11. V. N. Poddubny, Khimstroy, No. 3, 1933, 2125–2127.
12. S. M. Krugly, I. L. Lomakin and F. V. Iskritsky, Author's Certificate No. 85187, Bulletin of Inventions No. 9, 1950.
13. A. G. Sukasian, Author's Certificate No. 180172, Bulletin of Inventions No. 7, 1964.

14. P. A. Karnaushenko and N. N. Shvetzov, Author's Certificate No. 204325, Bulletin of Inventions No. 22, 1967.
15. N. N. Shvetzov, I. A. Bondarenko, O. P. Komlichenko and E. L. Baskin, Author's Certificate No. 434982, Bulletin of Inventions No. 25, 1974.
16. E. J. Peters and W. P. Zeman, Patent CA 1071120, A 19751222, 1980.
17. V. R. Pludek, Design and Corrosion Control, MacMillan Press Ltd., London. 1977, 383p.
18. V. S. Ivanov and F. Z. Serebriansky, Gas-oil Equipment of Generators with Hydrogen Cooling (in Russian), Energia, Moscow, 1980, 320p.
19. V. S. Ivanov, A. B. Goldshtein, L. I. Korneev and F. Z. Serebriansky, Author's Certificate No. 453350, Bulletin of Inventions No. 46, 1974.
20. D. M. Antony, S. J. Thrope and D. W. Kirk, Proceedings of Electrochemical Soc., University of Toronto, 1997, pp. 53–66.
21. L. M. Yakimenko, I. D. Modilevskaya and Z. A. Tkachek, Electrolysis of Water (in Russian), Khimiya, Moscow, 1970, 263p.
22. I. V. Riskin, M. I. Kadraliev, G, P. Tutaev et al., Author's Certificate No. 514152, Bulletin of Inventions No. 18, 1976.
23. E. E. Millway, M. H. Kleinmann, Corrosion, Vol. 23, No. 4, 1967, 88–97.
24. J. B. Cotton and H. Bradley, Chemistry and Industry, No. 2, 1958, 640–646.
25. M. I. Pasmanic, B. A. Sass-Tisovsky and L. M. Yakimenko, Producing of Chlorine and Caustic Soda, Handbook (in Russian), Khimiya, Moscow, 1968, 308p.
26. I. V. Riskin and Ya. B. Skuratnik, Zashchita Metallov, Vol. 20, No. 1, 1984, 97–102.
27. E. I. Dizenko, V. F. Novosiolov, P. I. Tugunov and V. A. Yufin, Anticorrosion Protection of Piping and Vessels (in Russian), Nedra, Moscow, 1978, 199p.
28. L. M. Yakimenko, Electrolyzers with Hard Cathode (in Russian), Khimiya, Moscow, 1966, 300p.
29. I. V. Riskin, Ya. B. Skuratnik, L. M. Lukatsky et al., Zashchita Metallov, Vol. 19, No. 6, 1983, 899–908.
30. A. N. Frumkin, Jurn. Fiz. Khimii, Vol. 23, No. 12, 1949, 1477–1482.
31. W. V. Baeckmann, W. Schwenk and W. Prinz (authors and eds.), Handbook of Cathodic Protection: Theory and Practice of Electrochemical Protection Processes, 3rd Edn, Gulf Publishing Co., Houston, 1997, 520p.
32. V. S. Kuzub, Anodic Protection of Metals against Corrosion (in Russian), Khimiya, Moscow, 1983, 184p.
33. A. I. Shtrimov, I. V. Zvolinskaya and I. S. Dobrodeev, Author's Certificate No. 99748, Bulletin of Inventions No. 1, 1955.
34. R. Juchniewich, W. Sokolsky and S. Sadousky, Patent No. 110807, Poland, 1977.
35. V. Bartel, J. Bistriansky and P. Novak, Author's Certificates No. 203685 and No. 203686, Czechoslovakia, 1979.
36. I. V. Riskin, V. B. Torshin and V. A. Timonin, Author's Certificate No. 1306979, Bulletin of Inventions No. 16, 1987.

CHAPTER 10

New Principles of Protection of Passive Metals Against Electrocorrosion in Electrochemical Plants

Contents

10.1. Protection of Metals With the Relationship $\Delta = E_{ox} - E_A < 0$, Against Corrosion Attack by an External Anodic Current, With the Help of Dimensionally Stable – Current Leak-Off Anodes

10.1.1. Theoretical basis

The results of the investigations of the corrosion stability of metals with different electrochemical characteristics considered in the previous chapters, point to the possibility of retaining the stability of a metallic structure in the field of an external current, if the metal possesses a high electrochemical activity. In this case, the term "electrochemical activity" means the ability of the metal to participate in oxidizing processes of the solution components on its surface, along with maintaining its passive state under the action of an anodic external current. Such a possibility was confirmed, particularly, when investigations were carried out on a Ti–2% Ni alloy in a chlor-alkali solution and of stainless steel 18-10 under the conditions of sodium perborate production. These metals possess a high electrochemical activity related to the oxidation of hydroxyl ions from the solutions. Anodic activation of the metals does not occur during this process, because the oxidation potential of hydroxyl ions is more negative than the anodic activation potentials of these metals, i.e., the potential difference:

$$\Delta = (E_{ox} - E_a) < 0 \qquad (10.1)$$

The part of a structure made of a metal that possesses the indicated properties, works as an anode. Electrochemical processes of the oxidation of the solution components take place on its surface under the action of the external anodic current. As shown, under the conditions of perborate production, 7 cm long and 5 cm diameter tube branches of stainless steel 18-10 are able to retain their stability at such high current magnitudes as 1.6 A. At the comparatively small dimensions of this part of the structure, the irregularity of the current and distribution of the potential over the metal surface can be ignored and its dimensions can be defined on the basis of the permissible magnitude of the anodic current density under the considered conditions.

It was shown in Chapter 6 that titanium possesses an unusually high activation potential, which is much higher than the oxidation potential of the solution components, such as chlorine and hydroxyl ions, i.e., the stability condition 10.1 is satisfied for titanium. Nevertheless, as opposed to the metals considered in the examples given above, titanium undergoes corrosion damage under the attack of low density external anodic currents, as was shown in Chapters 4 and 6. The barrier oxide layer formed on its surface deprives it of the electrochemical activity (ability to oxidize chlor–ions and other components of the solution). In other words, titanium is not able to work as an electrochemically active anode under the action of an external anodic current.

The fact that the activation potential of titanium is more positive than the oxidation potential of the solution components raises the possibility of stopping the "falling off" of the potential of the metal to the value of the activation potential caused by the attack of the external anodic current [1]. This idea is realized by setting up an electric contact to the areas of the titanium structure to be protected, with anodes that possess electrochemical activity and corrosion stability in the given aggressive environment. In particular, thermodynamically stable metals such as platinum and platinum group metals can be used as anodes (see Chapter 1, Figure 1.5). As the external anodic current attacks the metallic structure protected by these anodes, the titanium potential shifts to the positive side, up to the value of the oxidation potential of some of the solution components. Since the overvoltage of the oxidation processes on electrochemically active metals is comparatively low (at least, with respect to titanium), the entire current will focus on the anodes and will be spent for the oxidation reactions of the solution components. Realization of this idea involves the selection of optimal materials, constructions of the anodes, and optimization of their location on the titanium structure to be protected.

In electrochemical plants, current densities on the anodes of the modern electrolyzers attain very high values. For example, in chlor-alkali electrolysis plants, the current densities attain $100\,kA/m^2$ and more. Therewith, the anodes have to maintain a long-time stability and efficiency, which is characterized by a low overvoltage of chlor-ion oxidation to molecular chlorine, in accordance with the reaction:

$$2Cl^- = Cl_2 + 2e^- \tag{10.2}$$

The sizes of the leakage currents are several orders of magnitude lower than the sizes of the industrial electrolysis current. Hence, anodes of a small surface area are able to provide efficient corrosion protection of large-size titanium structures under the attack by leakage currents.

Graphite was used for a long time as a major anode material in diaphragm electrolyzers of chlor-alkali electrolysis plants [2, 3]. However, its stability was insufficient, owing to oxygen evolution on the anode surface, along with chlorine evolution. This oxygen oxidized the graphite that was accompanied by the formation of carbon oxides, and led to anode wear that, in turn, led to the distance between the anodes increasing, to the voltage in the baths increasing and, in the final analysis, to significant energy losses.

Owing to their thermodynamic stability, platinum and platinum group metals possess a high corrosion stability. Moreover, these metals are characterized by a low overvoltage of chlorine evolution at high current densities [2–6]. Since these materials are expensive, they are applied in industrial anodes in the form of thin coatings, in the order of magnitude of microns, applied to a titanium base. Nevertheless, the wear of these coatings leads to significant losses of the precious metals. In this connection, researches on less expensive anode materials are constantly being carried out all over the world. The possibilities of producing electrochemically active coatings of base metal oxides – Mg, Co, Pb, etc. – have been studied [2, 4, 7–9].

In the 1970's, titanium anodes with a titanium–ruthenium dioxide coating found extensive application in chlor-alkali electrolysis plants [2, 10–13]. Since these anodes are stable and do not change their dimensions during their operation, they received the name "dimension stable anodes" (DSA). They are less expensive than platinum coated anodes and have similar electrochemical characteristics.

The efficiency of the protection provided by DSA, or other anodes, depends on the overvoltage of the process of oxidation of the solution components on its surface in the given electrolyte and on the activation potential of the protected metal. This is illustrated by the diagram shown in Figure 10.1, in which simplified polarization curves are presented for anodes (1–3) and for metals (4, 5) which have activation potentials Ea_4 and Ea_5 and are protected by these anodes.

It is seen from Figure 10.1 that the higher (more positive) is the E_a value of the protected metal, the higher will be the oxidation current density of the solution components i_{ox} reached on the anode surface in the field of potentials which are more negative than the activation potentials E_a. In Figure 10.1, the maximal current density obtained for metal 4, which has an activation potential of E_{a_4}, is i'_{ox} and the maximal current density are obtained for metal 5, which has an activation potential of E_{a_5}, is i''_{ox}. It is seen that $E_{a_5} > E_{a_4}$ and, other things being

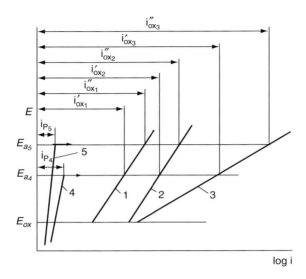

Figure 10.1   The dependence of the oxidation currents i_{ox} on the character of the polarization plots (1–3) and on the values of the activation potentials of the protected metals (4, 5).

equal, $i''_{ox} > i'_{ox}$. It is clear that the current density of the oxidation also directly depends on the oxidation overvoltage of the solution compounds on the anode, i.e., it depends on the slope of the polarization plot of the anode and on its shifting to negative values of the potentials (compare plots 1–3).

Thus, the protected metal must possess a high value of the activation potential E_a and the anodes must possess a high electrochemical activity to afford an effective protection from corrosion by an external anodic current. As these conditions are satisfied, the major part of the external current which reaches the protected structure is concentrated on the anodes, and it is spent not for the corrosion destruction of the metal, but for the oxidation reactions of the solution components on the anode surface.

Aqueous chloride and sulfate solutions belong to the most widely used media in electrochemical plants. Consequently, the oxidation of hydroxyl- and chlor-ions to molecular oxygen and chlorine, and the oxidation of the sulfate ion to persulfate are the most widely met reactions. The standard potentials of these reactions are quite high: 1.29, 1.36 and 2.01 V, respectively [14]. Among the structural metallic materials considered above, only titanium and its alloys have E_a values that are significantly higher in neutral, acid and alkali media than the above-mentioned values of the standard potentials.

10.1.2.   Experimental verification of the protection principle

To estimate the feasibility of the regarded method for the protection of titanium from corrosion by an external anodic current, anodic polarization plots were obtained on titanium and on anodes of graphite and of titanium–ruthenium dioxide coating (DSA), in 1 and 300 g/l NaCl solutions, respectively, at a temperature of 90°C (Figure 10.2).

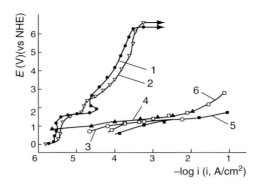

Figure 10.2 Potentiodynamic (0.72 V/h) anodic polarization plots on titanium (1, 2), and on anodes of graphite MG (3, 4) and of titanium with titanium-ruthenium dioxide coating (5, 6) in NaCl solutions: 300 g/l (1, 3, 5) and 1 g/l (2, 4, 6). Temperature, 90°C.

At a potential of 1.4 V, in the Taffel area of chlorine and oxygen evolution, the current density on the anode materials is 3–4 orders of magnitude higher than that on titanium, which at this potential has a stable passive state. It was shown above that under potentiostatic polarization, the current density on titanium is additionally reduced by an order of magnitude, due to the barrier film growth on its surface. The potentials at which the evolution of chlorine and oxygen start are about 4.5 V more negative than the activation potential of titanium ($\Delta = E_{ox} - E_a \approx -4.5$). This indicates that, in accordance with the considered principle, it is possible to implement an effective protection of titanium from corrosion by an external anodic current.

The current share which is spent on the anodic materials for oxygen evolution increases in dilute solutions [15]. Since the overvoltage of oxygen evolution on these materials is usually significantly higher than the overvoltage of chlorine evolution (see Chapter 1), the efficiency of these materials in the dilute solutions is noticeably reduced. This can be seen, for example, by comparing polarization curves obtained on the DSA anodes in 300 g/l and 1 g/l NaCl solutions (curves 5 and 6). Nevertheless, due to the high E_a value of titanium, these anodic materials can still provide effective protection of titanium. However, the possibility of elevated wear of the anodes related to the preferential oxygen evolution in the dilute solutions has to be taken into account. This effect is especially significant on the graphite anodes, due to their oxidation by the evolving oxygen [3, 15]; on the DSA anodes, the elevated coating wear occurs as a result of reaching the "critical potential" of the stability of the active coating [16]. Therefore, it is recommended that, along with the anodes, a means of reducing the current also be applied when the protection is carried out in dilute solutions.

Verification of the effectiveness of the protection principle was carried out on combined specimens made of titanium in contact with DSA anodes. Spade-shaped specimens of 0.15 mm thickness and 7.5 cm^2 working surface area, cut from titanium foil, were used for this purpose. Narrow strips of titanium foil with a titanium–ruthenium dioxide coating were welded to these specimens by resistance welding. The coating was made in full accord with the technology developed for DSA anodes [11]. The surface areas of the strips were equal to 0.5 and 0.3 cm^2, so the ratios of the titanium/DSA surface areas were equal to 15:1 and 25:1.

Under anodic polarization in a 1 g/l NaCl solution at a current density of 1 mA/cm^2 (calculated with respect to the total surface area of the combined specimens), potentials of 1.7 and 2.4 V, respectively, were stabilized on the combined specimens over half an hour, i.e., the maximal potential value remained about 4 V more negative than the titanium activation potential E_a. Therewith, the titanium part of the specimen did not undergo corrosion damage and maintained its initial metallic shine after 5 h of polarization. It is seen from Figure 10.2 that the potential values obtained on the combined specimens are similar to the values of the potentials installed on the DSA anode, assuming that the current is entirely concentrated on the surface of the anode. Thus, the studied anode, as an object of current leaking off, provides effective protection of titanium from corrosion by an external anodic current.

It was of interest to estimate the possibility of protecting systems which are characterized by the relationship $\Delta \approx 0$ using the principle under consideration. It was shown in Section 5.2.1 that carbon steel in alkaline solutions belongs to such systems, and it was noted that nickel and nickel coatings possess high corrosion stability and high electrochemical activity in these solutions.

As can be seen from Figure 10.3, in 1 N NaOH solution, at a temperature of 90°C, the overvoltage of the oxygen evolution on nickel and on steel with a nickel coating, is only 70 mV lower than that on carbon steel. However, as opposed to carbon steel, nickel and nickel coatings do not corrode in the field of oxygen evolution.

The assessment of the efficiency of nickel anodes in protecting carbon steel was carried out on combined cylindrical specimens of carbon steel in contact with nickel, that had a diameter of 10 mm and a steel to nickel surface area ratio of 5:1.

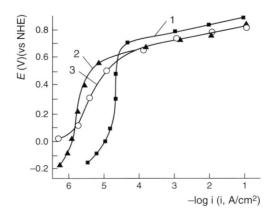

Figure 10.3 Anodic polarization plots in 1 N NaOH solution on: 1 – carbon steel St 3; 2 – carbon steel with nickel coating; 3 – nickel. Temperature, 90°C.

Connection of the elements of the combined specimens was carried out in a similar manner to the connection described in the previous chapter (where testing of the combined electrodes, which included dissolving the anodes for drainage of the external anodic current, was discussed). At a current density of $50\,mA/cm^2$ (calculated as the relationship between the impressed current and the total surface area of the combined specimen), the corrosion rate of the carbon steel was equal to $0.6\,g/m^2\,h$ and the corrosion rate of the nickel was close to zero. For a carbon steel specimen tested without contact with nickel, the corrosion rate was almost double: $1.1\,g/m^2\,h$. In accordance with the calculation based on the polarization curves, the current density on the carbon steel specimen was $\sim20\,mA/cm^2$, and on the nickel specimen it was an order of magnitude higher: $\sim200\,mA/cm^2$. Thus, the anode dissolution of the carbon steel was reduced, but not eliminated. The corrosion rate of the carbon steel specimens, measured without any contact with other metals at a current density of $20\,mA/cm^2$ (see Section 5.1.3), was close to the value obtained here. Thus, when $\Delta \approx 0$, i.e., at close values of the metal activation potential, E_a and the oxidation potential E_{ox} of the solution compounds on the anode surface, contact of the protected metal with the anode provides only partial protection of the metal from corrosion by an external anodic current.

The major advantage of the considered protection principle arises from the fact that it does not require the use of any external current sources (as in the case of cathodic protection by an impressed current and electrodrainage) and does not require the use of sacrificial anodes (as in the case cathodic protection by protectors and by dissolving anodes draining off the external current).

The anodes that are used for metal protection in accordance with the regarded principle concentrate the entire external current, or a major part of it, on their surface. Since this current is spent not for corrosion destruction of the metal, but for oxidation reactions of the solution components, these anodes received the name "current leak–off" anodes.

10.2. CORROSION PROTECTION WITH THE HELP OF DIMENSIONALLY STABLE ANODES, ORIENTED ALONG THE FIELD OF THE EXTERNAL CURRENT

10.2.1. Theoretical basis

The feasibility of the principle of corrosion protection discussed in the previous section are limited by the requirement $\Delta = (E_{ox} - E_a) < 0$ (10.1), which is imposed upon the structural materials. In practice, the only structural metallic materials that satisfy this requirement are titanium and its alloys. Metals such as niobium and tantalum, which have very high values of activation potential, are very expensive, and are used as structural materials only in singular cases.

Thus, it was necessary to find new ways of metal protection under the attack of external currents, which are not limited by the requirement $\Delta < 0$.

It was already emphasized that the corrosion protection of passive metals can be retained only under the conditions at which the potential value at the protected area of the structure is maintained in the field of the metal passive state. It is most probable that these conditions can be achieved by using electrochemical methods of protection.

When the dissolving anode is oriented along the field of the external current, its corrosion rate, maximal at its end, decreases in the opposite direction to the current flow direction, due to the irregular current, and potential distribution along the anode. This effect is used in applied electrochemistry, particularly for sharpening needles by electrolytic etching [17, 18].

It was shown in Chapter 8 that irregular currents and potential distributions also take place along the elements of structures made of passive metals and oriented along the field of the external current. In a cross-section in which the current flow through the metal is interrupted, the current and potential values are maximal. As the distance from the current interruption cross-section increases, in the direction opposite to the direction of the flow of the current, the current and potential values decrease: the current approaches zero and the potential approaches the value of the open circuit potential.

The regularities of the currents and potential distributions along tube-shaped elements of metallic structures that were considered in Chapter 8 made it possible to estimate the corrosion stability of these elements in the field of the external current, based on the condition $E_{max} < E_a$. When this condition is met, the structural element can be regarded as a corrosion–stable anode. It was shown however, that the metal of the estimated structure is often far from possessing the necessary combination of corrosion and electrochemical characteristics that afford the fulfillment of this condition.

The requirement $E_{max} < E_a$ can be reached by installing stable anodes at the areas of attack by external currents, which are oriented along the field of the external current and have the necessary length [19]. The effect of irregular potential and current distributions along these anodes in the field of the external current can be utilized in this manner for corrosion protection of the metallic structure.

There is no need to impose the requirement $\Delta < 0$ upon the protected structural metal when such anodes are applied. At each Δ value, by varying the anode length and the remote distance from the structure along the direction of the external current (taking into account E_a and other electrochemical and corrosion characteristics of the metal), it is possible, in principle, to achieve a reduction of the potential at the protected area to the necessary safe value.

As the principle of protection from electrocorrosion under consideration is implemented, the value of the oxidation potential of the anode E_{ox} does not depend on determining the characteristics of its interaction with the metal–aggressive environment. Generally, the maximum potential value at the anode end, which is at a maximum distance from the protected area of the structure along the field of the external current, can be higher, lower or equal to E_a. When the requirement $E_a < E_{max}$ is not fulfilled at the remote anode end, reduction of the potential takes place along the anode, towards the protected structure. The size of this reduction depends, first of all, on the dimensions of the anode. As shown in the previous chapter, the rate of the potential reduction along the anode depends on its polarization characteristics in the given aggressive medium, on the conductivity of the medium, on the geometric parameters of the anode and the protected structure, and on the magnitude of the external current [20].

The stable anode which is installed in the protected area of the piping or the tube-shaped structure, should preferably also be manufactured in the form of a tube. It touches the protected area as if it were a continuation of this area. The field of the external current is extended inside and along the tube; and the anode forms a part of this tube. With such a shape, the anode is always oriented along the field of the external current.

The equations of the potential distribution along a tube in the field of an external current I_e, which include all the above-mentioned characteristics of the metal and of the electrolyte, and also take into account the distribution of the potential in the electrolyte along the radius, are given in Chapter 8. These equations can be used in the same way to estimate the corrosion stability of a tube of a known length and for determining the length of the stable anode which is necessary to reduce the potential along this anode from some maximal value E_{max} at one of its ends, to a safe value $E < E_a$ at its other end [21]. The difference exists in only one detail: in the first case, the distribution of the potential along a homogeneous tube is considered, and in the second case, this distribution is along a combined tube (which includes the protected structure area CA and the anode BD) (Figure 10.4). In cross-section A of the point of their contact, the potential values of both sections are equal, but the current density changes abruptly, in accordance with the polarization characteristics of the protected metal and of the anode.

The fulfillment of the requirement $E_{max} < E_a$ can be reached not only by using an anode that is extended to a necessary length, but also by its remote location at the necessary distance along the direction of the external current. The second option makes it possible to reduce the anode length, and this can be realized by installing a tube AB made of conductive, (Figure 10.4b), or insulating, (Figure 10.4c), material

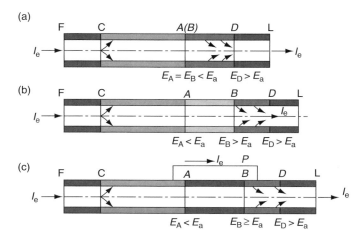

Figure 10.4 Protection from corrosion by the attack of an external current on tubes filled with electrolyte by using anodes: a – extended; b, c – remote, placed towards the direction of the external current field. BD – anode; AB – interstitial tube of conductive (b) or insulating (c) materials; FC and DL – parts of insulating tubes; P – external conductor (see color plate 6).

between the anode and the protected structure. Using a conductive material for tube AB is admissible in the case that the requirement $\Delta = E_{ox} - E_a < 0$ is satisfied for this material (see above).

The rate of the voltage drop along the conductive tube may be lower than that along an insulating tube, since the metallic tube partially "shunts" the electrolyte, due to leaking-off of a share of the current from its internal surface.

The change in the potential along the tube of an insulating material has a linear character, in accordance with Ω's law.

10.2.2. Investigation of the protection principle

The possibility of implementing of the considered protection principle was studied on stainless steel 18-10 in the acid–sulfate solution of the copper electrorefining process. It was shown previously (see Section 5.2.3) that under anodic polarization in this solution, steel 18-10 attains a transpassive state at potential values 1.1–1.2 V, which are more negative than the oxygen evolution potential. Thus, steel 18-10 is characterized under these conditions by the relationship $\Delta > 0$.

It will be shown further that anodes of lead possess satisfactory corrosion stability and high electrochemical activity in relation to the anodic oxygen evolution in the acid–sulfate solutions. The processes of anodic oxidation of the solution components on lead run at potentials above 1.7 V [22], which are significantly more positive than the activation potential of stainless steel 18-10.

The studies were done on a model of a compound tube (Figure 10.5) that consisted of a set of short tubes, 1, of steel 18-10. A similar model is described in work [23], in which it was applied for studies of the current distribution in the process of corrosion protection by sacrificial anodes. All the short tubes were insulated from one another by Teflon gaskets, 2. This made it possible to measure the current flowing between the adjacent short tubes by means of a milliammeter, 3. Capillaries of electrolytic bridges, 4, for measuring the potential were installed inside the tubes and were sealed by rubber gaskets, 5. The field of the external current for the compound bipolar tube was produced with the help of two tube-shaped

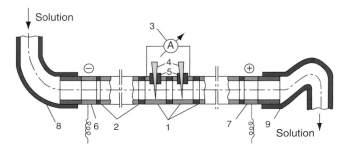

Figure 10.5 Model of a compound tube for estimation of corrosion protection efficiency by anodes oriented in the field of the external current (explanations are given in the text) (see color plate 7).

electrodes: a cathode, 6, of steel 18-10 and an anode, 7, of lead. The electrodes were connected to a controlled power source (a potentiostat that worked in a galvanostatic regime), which is not shown in the figure. In one part of the experiments, one of the poles of the potentiostat was connected directly to the short steel 18-10 tube. In this case the compound tube worked as a model of a monopolar tube. The interstitial tube (for remote placement of the anode) was made of titanium. The internal diameter of all the tubes, including the anode and cathode, was equal to 16 mm; the total length of the compound tube was equal to 1.6 m.

The duration of the tests was 1 h. A steady distribution of the potential was attained during this time in all areas of the compound tube.

The solution passed through the compound tube at a flow rate of 1 l/h. From a 5 l-volume vessel the solution flowed by gravity through rubber tube, 8, the compound tube and bend, 9, which ensured complete filling of all the tubes with the solution. Vessels for feeding the tubes and for reception of the passed solution are not shown in the figure.

Parallel to the experiments, the distribution of the potential was calculated along a compound specimen that was installed at the anodic end of the steel 18-10 tube and consisted of a 94 cm long titanium tube and a 6 cm long tube-shaped anode of lead.

Anodic polarization curves obtained on the titanium and lead, and the cathodic polarization curve obtained on the copper in the copper electrorefining electrolyte at room temperature, were used for the calculation (Figure 10.6). The equation for the cathodic curve was introduced into the calculation to allow for the influence of the cathodic end of the bipolar tube, where copper deposition takes place.

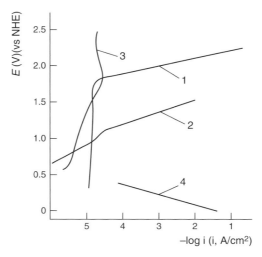

Figure 10.6 Anodic polarization plots on lead (1), stainless steel 18-10 (2) and titanium (3) and cathodic polarization plot on copper (4) in the copper electrorefining electrolyte. Temperature, 20°C.

The potential E_a (electric breakdown) of titanium under the considered conditions is above 140 V. It is seen in Figure 10.6 that at a potential value close to 2.5 V, the current density on the lead is about four orders of magnitude higher than that on titanium. Therefore, the activation potential on titanium is not attainable and the current will be completely concentrated on the lead anode.

The results of the tests are presented in Figure 10.7 and in Table 10.1. The model of the compound tube on which the measurements and calculations were carried out is as per the design given in Figure 10.4b. The letters marking the cross-sections at which the potential values are indicated in Table 10.1 are also used in accordance with this figure. When the model of the monopolar tube was studied, the steel 18–10 tube was used directly as a cathode.

As seen from Figure 10.7, the experimental and calculated curves of the distribution of the potential along the compound tube are close to each other. Moreover, from Figure 10.7 and from Table 10.1, it is seen that the potential values reached on the lead anode are significantly higher than 2 V, i.e., they are much higher than the transpassivation potential of the stainless steel. Nevertheless, due to the remote placement of the anode towards the direction of the external current (by installation of an interstitial titanium pipe, or by the use of an anode of the necessary length) the potential at the adjacent anodic end of the stainless-steel pipe is reduced to a value that is below the transpassivation potential value of the stainless steel.

Thus, from the data of Table 10.1, No. 1, it can be concluded that complete protection of the bipolar 18–10 steel tube was achieved when a 3 cm long anode was remotely placed at a distance of 40 cm (including the length of the anode itself). In this case, the maximum potential value on the stainless steel tube, $E_{max} = 0.83$ V, was lower than the activation potential E_a. When the magnitude of the external current was equal to 50 mA, only 32 mA entered the pipe and the rest of the current flowed through the electrolyte, in accordance with the results of the analysis of a bipolar tube presented in the previous chapter.

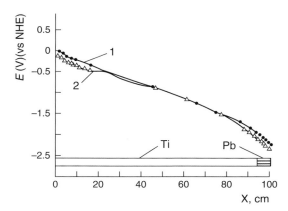

Figure 10.7 Experimental (1) and calculated (2) curves of potential distribution along a compound tube consisting of two parts: titanium and lead (anode) in electrolyte of copper electrorefining at an external current of 50 mA. Temperature, 20°C.

Table 10.1 Potential E and current I distribution along L long tubes of models of bipolar (BP) and of monopolar (M) tubes and corrosion rate C of tubes of steel 18-10

No.	Type of tube	I_e (mA)	Potentials (E) in cross-sections (V)			I_{max} inside the metal (mA)	Current I leaking off into the solution from separate short tubes						C (mm/ year)
							Steel 18-10		Titanium		Lead		
			D	B	A		L (cm)	I (mA)	L (cm)	I (mA)	L (cm)	I (mA)	
1	BP	50	2.3	2.06	0.83	32	20	32*	37	1	3	31	<0.1
2	M	100	2.45	2.2	1.17	100	9	4	97	2	3	94	0.5
3	M	100	2.33	2.1	0.93	100	9	1	147	4	3	95	<0.1
4	M	100	2.29	2.0	0.68	100	9	0.5	94	5.5	6	94	<0.1

* cathodic current

In the tests of a monopolar tube (No. 2), remote placement of the 3 cm long anode at a distance of 1 m, did not provide complete protection of 18-10 steel ($E_{max} = 1.17\,V > E_a$).

The average corrosion rate of the steel (without taking into account the irregularity of the corrosion distribution along the tube) was equal to 0.5 mm/year. Complete protection was reached either by increasing the distance of the remote placement to 1.5 m (No. 3) or by increasing the anode length from 3 to 6 cm (No. 4).

In accordance with the results of the current distribution measurements along the specimen in the form of a monopolar tube, 94–97% of the current leaked off from the anode and only 2–5.5% leaked off from the titanium pipe, which maintained a passive state under these conditions. When complete protection of the stainless steel was reached, the share of the current that leaked off from the stainless-steel tube did not exceed 1%. When the anode length or the distance of its remote placement was insufficient (No. 2), the value of the anodic current on the stainless-steel tube increased to 4%.

The obtained results prove the possibility of metal protection from electrocorrosion by anodes oriented in the field of the external current, when a sufficient length or distance of their remote placement are provided along the field of the external current.

It was shown in the previous chapter that the application of nickel anodes – current leak-offs – did not provide complete electrocorrosion protection of structural elements made of carbon steel in alkaline solutions. As opposed to this, using nickel anodes oriented in the field of an external current should provide complete protection of these structural elements.

10.3. PROTECTION OF METALS AGAINST CORROSION ATTACK BY AN EXTERNAL CATHODIC CURRENT

Although in the majority of cases electrocorrosion is the result of attack by external anodic currents, it was shown in Chapter 7 that metals such as titanium can undergo hydrogenation and corrosion by external cathodic currents and the corrosion rate under real conditions may attain high values.

The cathodic activation of a passive metal occurs as a result of the shift of the metal potential from its corrosion (open circuit) potential to the negative side, up to the cathodic activation potential E_{ca}. Prevention of the shift in the potential to a value which is more negative than the permissible value E_{max} ($E_{max} > E_{ca}$), i.e., complying with the condition (8.1b) affords protection from attack by an external cathodic current.

In some cases the protection of metals from corrosion attack by external cathodic currents can be similar to the protection by current leak-offs, which was considered in this chapter. Such protection is possible when the potential of the cathodic reduction of the solution components is more positive than the potential of the cathodic activation of the protected metal, $E_{red} > E_{ca}$, or the algebraic difference $\Delta_c = E_{red} - E_{ca} > 0$ (assuming that the cathodic potentials are negative with respect to the hydrogen electrode).

When the condition $(\Delta_c < 0)$ is not fulfilled, the protection from corrosion attack by a cathodic current can be reached with the help of stable cathodes, oriented along the field of the external current, as in the case of the protection by the stable anodes in the case of attack by an external anodic current. Under such conditions, the maximum negative potential is realized at the cathode end which is remote from the protected structure area, and an increase in the potential (shift to the positive side) takes place along the cathode, towards the protected structure area. Reduction of the solution components takes place at the remote cathode end and fulfillment of the condition (8.1b) is not obligatory. At a sufficient cathode length or at its sufficiently remote placement, the potential that is reached at the end of the protected object from the side of the cathode will be more positive than E_{ca}, and the protected metal will not undergo cathodic activation.

The distribution of the potential along a tube-shaped cathode can be calculated, in a similar manner to the calculation of the potential distribution along an anode, with the help of the equations given in Chapter 8. Therewith, the cathodic characteristics $i = f(E)$ for a cathode which is stable in the given electrolyte should be used for this calculation.

Studies of the possibility of cathodic protection with the help of a cathode oriented along the field of the cathodic current were carried out on an installation similar to the one presented in Figure 10.5 with a uniform titanium tube of internal diameter 16 mm. The experiments were carried out at room temperature in 1 M Na_2SO_4 solution acidified to pH 2 with sulfuric acid. The conductivity of the solution was $0.089 \, \Omega^{-1} cm^{-1}$.

The titanium tube was connected to the negative pole of a current source (monopolar tube). A short titanium tube with platinized internal surface was used as the anode. The external current was equal to 200 mA.

It is seen from Figure 10.8a that along a tube that has a length of 20 cm, a fast increase in the potential, of 1.5 V, takes place. The results of the measurements (curve 1) and of the

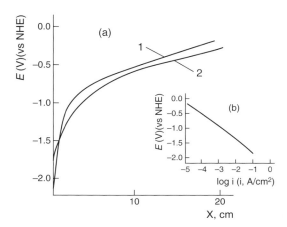

Figure 10.8 a – experimental (1) and calculated (2) potential distribution along a tube-shaped titanium cathode of 16 mm diameter in 1 M Na_2SO_4 solution acidified by sulfuric acid to pH 2. External current 200 mA, temperature 20°C; b – cathodic polarization plot on titanium in the same solution that was used for the calculation.

calculation (curve 2) are close to each other. The cathodic polarization curve obtained in the same solution (Figure 10.8b) was used for the calculation of the potential distribution.

Thus, when $E_{max} < E_{ca}$, the stable cathodes that are oriented along the field of the external current, and that have a sufficient length or are placed at a sufficiently remote distance are able to protect metallic structures from corrosion under the attack of external cathodic currents.

It was shown in Chapter 7 that in technological solutions containing chlorides, under the conditions of the action of cathodic current, the ends of the titanium pipes can play the role of stable cathodes, when the conditions can be produced for the accumulation of oxygen-chlorine compounds near the surface of these areas. These areas of the pipes can be considered as cathodes oriented along the field of the external cathodic current. The potential increases along these areas to values at which the hydrogenation and corrosion of titanium does not occur. Specific technical solutions for realizing such corrosion protection will be considered further.

Discharge of the hydroxonium ion, the major component of the aqueous electrolytes, on the metal surface is in many cases, a determining stage in the cathodic process. This discharge is characterized by the Taffel dependence of the overvoltage $\acute{\eta}$ of the cathodic reaction at the current density: $\acute{\eta} = a + b.\log i$ [24, 25]. The coefficient a in this equation is strongly dependent on the nature of the metal. According to the data in Ref. [25] it is equal for iron, nickel and titanium, respectively, to 0.7–0.76, 0.63–0.65 and 0.82–0.83, and for platinum and palladium, respectively, to 0.1–0.3 and 0.24–0.53 (the first values are for acid solutions, and the second, for alkaline solutions). The coefficient b varies in a much narrower range and, in accordance with the data in Refs. [24, 26], for most of the metals it is equal to 0.116. It is indicated in Ref. [25] that the coefficient b is significantly lower for metals that have a low hydrogen overvoltage than for other metals: for platinum and palladium it is equal, respectively, to 0.03–0.1 and 0.03–0.13, and for titanium it is equal to 0.14. At such relationships between the electrochemical characteristics, the current density of the cathodic reduction of hydroxonium must be higher on platinum than on titanium by no less than two orders of magnitude [27], other things being equal. Taking into account the high corrosion stability of platinum and palladium in most aggressive media, it may be expected that these metals can be used as effective cathodes for the protection of metals from attack by external cathodic currents. This is proved by the data in Ref. [28] on the reduction of titanium hydrogenation when it is in contact with palladium under the conditions of cathodic polarization in sulfuric acid solutions. It is clear that the efficiency of platinum in these conditions will be even higher. The restriction on the application of platinum as a stable cathode is related to its high price.

Platinized titanium cathodes are not stable, owing to scaling of the platinum coating by hydrogen, which penetrates under the coating and forms a hydride layer [29]. An intermediate layer of ruthenium dioxide increases the stability of the cathode [29], but it does not solve the problem of the development of stable cathodes based on accessible and low-cost materials. The anodes with the ruthenium dioxide coatings and other types of anodes with coatings based on metal–oxide compounds, lose their stability as a result of variations of the external currents by magnitude and, especially, by direction [30–32].

Some progress was achieved during the last decades, in the development of ceramic anodes with a coating of titanium suboxides, having a general formula Ti_nO_{2n-1}, on the base [4]. These coatings possess a high conductivity and stability under the action of both anodic and cathodic currents and are used as stable cathodes in sulfuric acid and other media. The oxides Ti_4O_7 and Ti_5O_9 are most often used for producing these electrodes [33, 34].

The specific challenge in the application of the stable anodes and cathodes for implementation of the considered principles of the protection of metallic structures from electrocorrosion, requires finding solutions to the problems related with the selection and development of materials for producing these electrodes, or development of the means that would provide their stability.

The electrochemical characteristics and stabilities of the electrodes determine to a great extent the possibilities and the ways of their practical implementation for the developed methods and means of protection from electrocorrosion.

10.4. PROTECTION OF METALS AGAINST CORROSION ATTACK BY LEAKAGE CURRENTS THAT PERIODICALLY CHANGE THEIR DIRECTION

The protection of metals against corrosion by an external current becomes more complicated when the current can randomly change its direction. This may occur due to changes in the number of operating electrolyzers in relation to the zero point of a set and to the periodical inclusion of additional current sources to the circuits of the electrolyzers, etc. In modern plating plants, periodic reversal of a current is used for depositing a metal on a cathode. For example, the uniformity and quality of a cathodic deposit is raised by briefly changing the current direction (up to 7% of the time) throughout the entire plating cycle [35].

Titanium piping, which is able to maintain its corrosion resistance under the action of a cathodic current in chloride-containing media, corrodes under the attack of a comparatively low anodic current. On the other hand, the stability of most coatings on stable anodes is reduced under the attack of a comparatively low cathodic current [30–32].

In these cases, it is necessary to use a protective electrode which is effective for both anodic and cathodic currents. Ceramic electrodes, discussed in the previous section, have such properties. However, they are expensive and cannot be used in all aggressive media in the zones of action of the external currents.

Therefore, it was of interest to study a composite of a stable anode with a stable cathode and to find the conditions under which the combination would be effective in a protection system. Each of these electrodes should have a low overvoltage for its intended reaction and should protect not only the metal structure but also the other electrode.

Following is a necessary condition of such protection at any relative arrangement of the anode and cathode in the field of an external anodic current I_e:

$$I_a^{(a)} \gg I_a^{(c)} \tag{10.2}$$

and of the external cathodic current:

$$I_c^{(c)} \gg I_c^{(a)}, \tag{10.3}$$

where the sub- and superscripts of I, like in the following equations, designate: subscript – direction of the current affecting the electrode; superscript – the type of electrode (i.e., a – anode, c – cathode) through which the current passes.

The conditions (10.2) and (10.3) indicate that the system will be operational when the anode and cathode of the composite electrode can accept a large part of the "like-named" current, i.e., anodic for the anode and cathodic for the cathode. Consequently, they relieve the action of the "unlike-named" current on the opposite parts of the composite electrode.

Condition (10.2) (anodic polarization) is easily satisfied. It is enough to use a stable anodically active material as the anode (e.g., titanium with a ruthenium dioxide coating in chloride solutions) and as the cathode, a metal that is passive at anodic polarization and has a high anodic activation potential. The latter requirement is satisfied by titanium, in chloride and sulfate solutions. Achievement of condition (10.3), even when the anode and cathode are similar in their cathodic activity, is possible by using the above-considered effect of changing the potential along the electrode placed along the field of the external current. In accordance with this effect, taking into account the non-uniformity of the potential distribution along the electrode in the field of the external current, the anode and cathode must be disposed in a sequence where the anode is protected by the cathode. This is done by placing the anode between the cathode and the area of the metal structure to be protected (Figure 10.9a). In such a sequence the direction from the anode to the cathode coincides with that of the external anodic current field. Thus, condition (10.3) governs the sequence of the electrode positioning [36].

Assuming that the anode and cathode of the composite electrode are resistant to their "like-named" current, it is possible to formulate the conditions to maintain the stability of the considered system based on the maximum permissible anodic

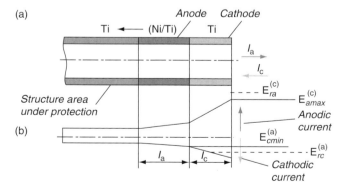

Figure 10.9 Potential distribution (b) along the composite electrode (a), placed along the field of external currents of anodic (I_a) or cathodic (I_c) directions (see color plate 8).

and cathodic potentials (taking the potential sign into account) produced by the "unlike-named" current (Figure 10.9b).

For the anodic-direction current we have:

$$E_{ra}^{(c)} > E_{a\,max}^{(c)} \tag{10.4}$$

where E_{ra}^{c} is the limiting permissible anodic activation potential for the cathode and $E_{a\,max}^{(c)}$ is the maximum cathodic potential that can be reached during the anodic polarization.

For the cathodic-direction current:

$$E_{rc}^{(a)} < E_{c\,min}^{(a)} \tag{10.5}$$

where $E_{rc}^{(a)}$ and $E_{c\,min}^{(a)}$ are, respectively, limiting permissible and minimum potentials at the anode during the cathodic polarization.

Condition (10.4) is satisfied if the cathode length $l^{(c)}$ is less than the limiting permissible length $l_{a\,max}^{(c)}$, i.e.,

$$l^{(c)} < l_{a\,max}^{(c)} \tag{10.6}$$

Condition (10.5) is satisfied when the cathode length $l^{(c)}$ is enough to shift the potential along the electrode in the positive direction to a value sufficient to prevent corrosion attack thereon. This length should be higher than the minimum permissible value:

$$l^{(c)} > l_{c\,min}^{(c)} \tag{10.7}$$

Combining inequalities (10.6) and (10.7) gives a condition which is sufficient for the system to resist a variable-direction external current:

$$l_{c\,min}^{(c)} < l^{(c)} < l_{a\,max}^{(c)} \tag{10.8}$$

This expression clearly shows the corrosion resistance of a given system to be governed by the cathode length, within certain limits.

Let us determine the relationship between $l_{c\,min}^{(c)}$ and $l_{a\,max}^{(c)}$ when the cathode is made of titanium. As shown in Chapter 6, titanium is characterized by a high anodic activation potential: over 5 V in chloride and over 100 V in sulfate solutions. In addition, the anodic polarizability of passive titanium is high, so that the potential along the electrode during the anodic polarization should vary more slowly than along a metal with a lower polarizability [20]. It is therefore postulated that when the length of the titanium cathode $l_{a\,max}^{(c)}$ is high enough, the potential at its free end $E_{a\,max}^{(c)}$ will not reach $E_{ra}^{(c)}$, i.e., condition (10.4) will be satisfied.

On the other hand, with a cathodic current, there is usually a small interval between the potential of cathodic reduction of the solution components on the cathode and the potential of initiation of the anode destruction (usually lying within the boundaries of 1–1.5 V). It may therefore be expected that the necessary shift of the potential to the positive along the electrode from its free end to a point safe for the anode, may be achieved with a short cathode. Hence, $l_{c\,min}^{(c)} < l_{a\,max}^{(c)}$ is possible, and there should be a range of $l_{a\,max}^{(c)} < l_{c\,min}^{(c)}$ where the required cathode length $l^{(c)}$ can be found.

Piping used in electrochemical plants can be protected by using tube-shaped electrodes of the same diameter as the pipes to be protected. The expressions considered above may be used to determine the permissible range of lengths of cathode and anode placed along the field of the external current.

Example. The possibility was assessed using a composite electrode to protect a titanium overflow device for withdrawing caustic liquor from a diaphragm electrolyzers into a header. The maximum measured leakage current from the marginal cells of a series reached 2 A without any change of direction. A change was possible only near the zero point of the set where the current was much lower. Assuming that the current can change direction in the middle section between the last cell and the zero point, we can estimate conservatively that the maximum possible anodic current in this section would be 1.3 A, with a cathodic component of 0.25 A. The tube diameter was 50 mm and the solution resistance 1.1 Ω cm. The anodic and cathodic polarization curves used for the computation are shown in Figure 10.10.

Titanium, anodically oxidized in a solution of 200 g/l Na_2SO_4 + 100 g/l NaOH at a temperature of 90°C and a current density of 5 mA/cm^2 for 1 h, was used as the cathode. The oxidation was carried out to prevent titanium hydrogenation during the cathodic polarization. According to previous data, anodic activation of titanium in caustic liquor occurs at a potential of 6.5 V. We assumed $E^{(c)}_{a\,max} = 6V$, which is somewhat lower than the activation potential.

The anode was also made on titanium base and had a Ni/Ti coating produced in an electric spark. This type of electrochemically active coating will be considered further. It was found that at potentials up to −0.8 V, the anode remained stable during the cathodic polarization, but at more negative potentials it was destroyed as a result of spalling of particles of the coat from its surface.

The maximum negative potential value at the free cathode end, at an external cathodic current of 0.25 A, was −1.3 V. Calculations have shown that the potential drops along the cathode towards the anode, from −1.3 to −0.8 V, when the

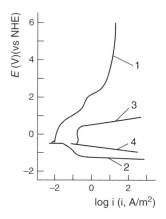

Figure 10.10 Potentiostatic polarization plots on titanium (1 and 2) and on Ni/Ti coating (3 and 4) in a solution of 120 g/l NaOH + 200 g/l NaCl. Temperature, 90° C. 1 and 3 – anodic plots; 2 and 4 – cathodic plots.

cathode length $l_{c\,min}^{(c)} = 55$ cm. The maximum permissible cathode length under anodic polarization $l_{a\,max}^{(c)} = 100$ cm. Thus, the permissible cathode length should be within the range 55 cm $< l^{(c)} < 100$ cm. If we take $l^{(c)} = 80$ cm, the required minimum calculated anode length would be 3.5 cm. These anode and cathode lengths in the composite electrode should provide long and reliable protection of the caustic liquor overflow device when subjected to action by currents with changing directions.

Thus, the composite electrode includes an anode and a cathode, which protect one another, and together they provide protection of metallic structures from electrocorrosion.

10.5. PROTECTION OF METALS AGAINST CORROSION ATTACK BY EXTERNAL CURRENTS AT THE DESIGN STAGE

Methods and means of metal protection from corrosion considered at the design stage of a project are discussed in many monographs and textbooks on metal corrosion [37, 38]. There are also monographs which are completely dedicated to this task [39, 40]. Most of these means are designed to prevent the development of local corrosion processes. Differential aeration and concentrations (stagnant zones, crevice effects) and galvanic couples (contact of different metals, preferential corrosion of welding seams) constitute the major part of these processes.

Corrosion by stray currents is regarded as the kind of corrosion that cannot be foreseen during the design. Consequently, the means for its prevention must have a universal character, which does not take into account direction, size and other characteristics of stray currents, since these data are not known. Isolation of the metallic structures by polymer coatings, electrical insulation of the stray-current sources from the ground, electrodrainage, cathodic protection, etc., which belong to these means, were discussed in Chapter 2.

The majority of the data used at the design stage for protection from corrosion by stray currents consists of the geometric parameters of the metallic structures to be protected and the electric resistance of the ground. Data on the latter parameter which are particularly necessary for the design of cathodic protection, are not usually exact owing to variations in the conductivity along the structure and at different seasons. The initial state of the metal (active or passive), its polarization, and other electrochemical characteristics, usually do not belong to the determining factors which should be taken into account when the protection from corrosion by stray currents is designed.

When the protection against electrocorrosion of metallic structures operating in aggressive environments of electrochemical plants is designed, external currents must be considered as a major factor of the environmental aggressiveness in the zones of their unavoidable action. In most cases, contrary to the situation with stray currents, the direction, order of magnitude and distribution character of leakage currents in electrochemical plants can be determined at the design stage.

The quantitative characteristics of the aggressive media, such as concentrations, temperatures, flow rates and conductivities, are strictly designated by the technology conditions and can also be used as parameters for the necessary calculations at the design stage.

Of course, the general means of protection, such as reducing current leakages from the electrolyzers and insulation of piping, which were considered in Chapters 2 and 9, must be applied in electrochemical plants, whenever possible. It was also noted in Chapter 9 that when electrodrainage does not lead to significant current losses, owing to metal hydrogenation by a cathodic current and to corrosion damage of the metallic equipment located outside the zones of the leakage current action, this method of protection from electrocorrosion can be successfully used. Consequently, it can be included in the project at the design stage.

All the considered results show that the structural metals used in the electrochemical plants are in a passive state, in the absence of external currents. Metal destruction under the action of external currents occurs as a result of a shift in the potential of the metal out of the boundaries of its passive field.

Protection of metals from electrocorrosion in electrochemical plants is mainly aimed at retaining the potential of the metal inside the passive field under the action of external currents. Such a restriction of the problem enabled the development of special electrochemical methods of protection of the passive metals in the field of external currents by using either dissolving or stable anodes and stable cathodes. The relationship between the metal activation potential E_a and the solution oxidation potential E_{ox} are taken into account with this type of protection. Using electrodes oriented along the field of the external current and electrodes remotely placed along this field, enlarged the possibilities of the corrosion protection. All the indicated methods and means of protection can be applied at the design stage, and all the necessary protection criteria can be calculated with the help of the equations given in Chapter 8.

It is worthwhile remembering that when the corrosion stability of metallic structures is estimated in the field of an external current, the electrochemical characteristics (including polarization curves) of the metal, as well as the characteristics of the aggressive environment, must be taken into account, along with the data on the geometric parameters of the metallic structure to be protected and on the direction and magnitude of the external current.

The selection of a metal as a material of construction for structures operating in the field of an external current can be done, in some cases, on the basis of its electrochemical characteristics in a given aggressive environment. When no barrier layer is formed on the metal surface and the metal has a fairly high electrochemical activity, it may be able to retain its corrosion resistance under the action of external currents without any additional means of protection. Such data were given in Chapter 8, in examples on the corrosion stability of a Ti–2% Ni alloy in a chloralkali solution and of steel 18-10 in a perborate solution. The estimation of the corrosion stability of the structures made of these metals at the design stage can be carried out with the help of the equations given in Chapter 8.

In addition, the equations given in Chapter 8 make it possible to determine the safe dimensions of the structural elements in the form of pipes, which can also be considered as a means of electrocorrosion protection at the design stage.

In some cases, for high values of the activation potential E_a (in the order of tens of volts and more), shifting to the E_a value can be prevented by sectionalization of the piping. The length of a pipe section is calculated at the design stage on the basis of the condition $E_{max} < E_a$. As a result of this sectionalization, the piping is able to maintain its stability in the field of an external current without using any additional means of protection from electrocorrosion. Insulation of a pipe from an external current source, i.e., transforming a monopolar tube into a bipolar tube (like insulation of the pipe sections in the case of sectionalization) also makes it possible, in some cases, to prevent a shift in the potential of the metal to the E_a value at the pipe end.

Along with the selection of metals with high E_a values, means must be provided for the prevention of a reduction of the E_a in the action zones of the external current. As shown in Chapter 6, a sharp drop of the E_a value of titanium can occur in crevices and on welding seams. One of the constructive solutions to exclude the welding seam from the action zone of the external current is to weld the flanges to the external tube surface. Such a solution was successful when a steel 18–10 pipe was used in the acid–sulfate copper electrorefining solution. It will be shown further that in the same environment, corrosion resistance of titanium degassing tanks was reached. The tanks were damaged at the welding zones on the bottom, which coincided with the action zones of an external anodic current. Welding titanium legs to the bottoms removed the bottoms from the action of the external currents.

Thus, the protection from electrocorrosion of metallic structures in the aggressive environments of electrochemical plants must be based on the knowledge, not only of the geometric parameters of the structure to be protected, but also of the direction and magnitude of the external currents and of the corrosion and electrochemical parameters of the structural metal and of the aggressive media. In contrast to the situation of protection from corrosion under attack by stray currents, all these data can be obtained and used in the design stage.

REFERENCES

1. I. V. Riskin, G. P. Tutaev, N. D. Tomashov et al., Author's Certificate No. 518983, Bulletin of Inventions No. 3, 1977.
2. L. M. Yakimenko, Electrode Materials in Applied Electrochemistry (in Russian), Khimiya, 1977, 264p.
3. F. I. Mulina, L. I. Krishtalik and A. T. Kolotukhin, Jurn. Prikladnoy Khimii, Vol. 38, No. 12, 1965, 2808–2827.
4. M.O. Coulter ed., Modern Chlor-Alkali Technology, Ellis Horwood Ltd, Chichester, Section B, Coatings for Metal Anodes and Cathodes, 1980, pp. 99–149.
5. L. M. Yakimenko, S. D. Khodkevich, E. K. Spasskaya et al., Platinum-Titanium Anodes in Applied Electrochemistry (in Russian), Review "Itogi Nauki i Tekhniki", Electrokhimiya, VINITI, Moscow, Vol. 20, 1982, pp. 112–152.
6. R. U. Bondar', V. S. Sorokendia and E. A. Kalinovsky, Dimensionally stable anodes with palladium oxide(in Russian). Proceedings of Fifth All-Union Conference "Dimensionally Stable Anodes and their Application in Electrochemical Processes", NIITEKhim, Moscow, 1984, p. 15.
7. Ya. M. Kolotirkin and D. M. Shub, Today State and Future of Investigations of Anode Materials of Oxides of Base Metals (in Russian), Review "Itogi Nauki i Tekhniki", Electrokhimiya, VINITI, Moscow, Vol. 20, 1982, pp. 3–43.

8. V. L. Mosckalevich, N. P. Juk and E. A. Kalinovsky, Dimensionally Stable Anodes for Aqueous Solutions Electrolysis (in Russion), Collected articles "Khlorine Producing Industry", NIITE-Khim, Moscow, Vol. 12, 1976, pp. 12–13.

9. E. Ya. Kriuchkova and G. N. Trusov, Zashchita Metallov, Vol. 19, No. 3, 1983, 442–444.

10. L. M. Elina, V. M. Gitneva, V. I. Bistrov and N. M. Shmigul', Electrokhimiya, Vol. 10, No. 7, 1974, 68–70.

11. V. N. Antonov, V. I. Bistrov, V. V. Avksentiev et al., Khim. Promishlennost, No. 8, 1974, 600–603.

12. A. F. Mazanko, V. L. Kubasov and V. B. Busse-Machukas, Application of metal-oxide anodes in producing of chlorine, caustic soda and oxygen-chlorine compounds. Proceedings of Fourth All-Union Conference. "Dimensionally Stable Anodes and their Application in Electrochemical Processes", NIITEKhim, Moscow, 1984, 3–5.

13. M. A. Warne, Materials Performance, Vol. 18, No. 8, 1979, 32–38.

14. A. M. Sukhotin, ed., Electrochemistry Handbook (in Russian), Khimia, Leningrad, 1981, 486p.

15. N. N. Bibikov, E. Ya. Liublinsky and L. V. Povarova, Electrochemical Protection of Sea-Going Ships against Corrosion (in Russian), Sudostroenie, Leningrad, 1971, 262p.

16. D. V. Kokoulina, Yu. I. Krasovitskaya and T. V. Ivanova, Electrokhimiya, Vol. 14, No. 3, 470–474.

17. V. A. Obukhov, Author's Certificate No. 265333, Bulletin of Inventions No. 10, 1970.

18. A. S. Loginov, Yu. E. Blok, I. V. Riskin et al., Author's Certificate No. 1235602, Bulletin of Inventions No. 21, 1986.

19. I. V. Riskin, V. A. Timonin, Ya. B. Skuratnik et al., Author's Certificate No. 934324, Bulletin of Inventions No. 14, 1983.

20. A. N. Frumkin, Jurn. Fiz. Khimii, Vol. 23, No. 12, 1949, 1477–1482.

21. I. V. Riskin, B. Skuratnik, K. M. Lukatsky et al., Zashchita Metallov, Vol. 19, No. 6, 1983, 899–908.

22. M. A. Dasoyan and I. A. Aguf, The Modern Theory of Accumulator (in Russian), Energiya, Leningrad, 1975, 312p.

23. I. A. Stepanov and M. M. Makarova, Zashchita Metallov, Vol. 4, No. 1, 1968, 21–26.

24. N. D. Tomashov, Theory of Corrosion and Protection of Metals, The Macmillan Company, New York, 1966, 672p.

25. L. I. Antropov, Theoretical Electrochemistry (in Russian), Visshaya Shkola, Moscow, 1965, 511p.

26. A. L. Rotinian, K. I. Tikhonov and I. A. Shoshina, Theoretical Electrochemistry (in Russian), Khimiya, Leningrad, 1981, 423p.

27. N. D. Tomashov and R. M. Altovsky, Corrosion and Protection of Titanium (in Russian), Mashinostroenie, Moscow, 1963, 168p.

28. N. D. Tomashov, V. N. Modestova and R. P. Vasilieva, Fiz.-Khim. Mekhanika Materialov, Vol. 4, No. 3, 1968, 346–351.

29. L. A. Mikhailova, S. D. Khodkevich, Ya. B. Skuratnik et al., Zashchita Metallov, Vol. 18, No. 3, 1982, 371–375.

30. A. A. Uzbekov, V. G. Lambrev and I. F. Yazikov, Electrokhimiya, Vol. 14, No. 8, 1978, 1150–1159.

31. L. I. Liamina, N. I. Korol'kova and K. M. Gorbunova, Electrokhimiya, Vol. 8, No. 5, 1972, 651–655.

32. V. S. Klementieva, Corrosion Behavior of Oxide-Ruthenium Anodes in Electrochemical Producing of Chlorine and Oxigen-Chlorine Compounds by Electrolysis of NaCl Solutions (in Russian), Khimiya, Moscow, 1984, 168p.

33. P. C. S. Haufield, Development of a New Material: Monolithic Ti_4O_7 Ebonex® Ceramic, (2nd Edn), Cambridge, Royal Soc. of Chemistry, 2002, 100p.

34. J. R. Smith, F. C. Walsh and R. L. Clarke, Journal of Applied Electrochemistry, Vol. 28, No. 10, 1998, 1021–1023.

35. A. I. Levin, Electrochemistry of Non-Ferrous Metals (in Russian), Metallurgiya, Moscow, 1982, 256p.

36. I. V. Riskin, Ya. B. Skuratnik and V. B. Torshin, Zashchita Metallov, Vol. 23, No. 2, 1987, 307–311.

37. I. Ya. Klinov, P. G. Udima, A. V. Molokanov and A. V. Goriainova, Corrosion Resistant Chemical Equipment, Handbook (in Russian), Mashinostroenie, Moscow, 1970, 591p.
38. K. Trethewey and J. Chamberlain, Corrosion for Science and Engineering, Langman, Harlow, 1995, 466p.
39. V. R. Pludek, Design and Corrosion Control, Macmillan Press Ltd, London, Basingstoke, 1977, 383p.
40. R. J. Landrum, Fundamentals of Designing for Corrosion Control: A Corrosion Aid for the Designer, NACE International, Houston, 1989, 352p.

CHAPTER 11

ELECTRODES FOR METAL PROTECTION AGAINST CORROSION ATTACK BY LEAKAGE CURRENTS IN ELECTROCHEMICAL PLANTS

Contents

The principles that were considered in the previous chapter enable electrochemical protection against attack by external currents with the help of stable anodes and cathodes, without using sources of currents and of dissolving electrodes. In the process of implementation of these principles in practice, the problems related to the selection and development of materials and to the design of stable anodes and cathodes are put in the forefront.

11.1. REQUIREMENTS UPON MATERIALS OF ELECTRODES THAT ARE USED FOR PROTECTION AGAINST CORROSION ATTACK BY EXTERNAL CURRENTS

The requirements demanded of the materials of the electrodes (anodes and cathodes) for commercial electrolysis and for protection against corrosion by external currents are similar in many respects [1]. The main ones are the following: high stability at a maximal current density, high electrochemical activity, sufficiently high mechanical strength, availability, low cost, adaptability and ease of manufacture.

At the same time, the different applications and operating features of the electrodes for protection against electrocorrosion leads to some significant differences in the requirements that are placed upon them. First of all, it is not necessary to limit the selection of materials for the electrodes by the requirement of low overvoltage of evolution of some of the solution components as is done in the case of commercial electrolysis. For example, anodes for the electrolysis of sodium chloride must have as low an overvoltage of chlorine evolution as possible and the highest possible overvoltage of the evolution of oxygen which contaminates the final product (chlorine). The sizes of the leakage currents are several orders of magnitude lower than the sizes of the commercial electrolysis currents. Therefore, the relationship between the quantities of the evolved chlorine and oxygen on the anode, which is used in the same plant for corrosion protection, are not important, since it does not noticeably influence the composition of the final product. Vice versa, it is preferable to use stable anodes that possess a low overvoltage of oxygen evolution, because such anodes are the most universal for the aqueous solutions. Their ability to provide corrosion protection does not depend on the composition of the solution, since the external current is spent for water decomposition. In this case, the requirements placed upon the anodes for protection against electrocorrosion are similar to the ones that are placed upon the anodes for cathodic protection by an impressed current. At the same time, for energy saving in both commercial electrolysis and cathodic protection, the anodes must have a minimal overvoltage. It is desirable to reduce the overvoltage in the systems of protection against corrosion by external currents since this increases the efficiency of this kind of protection.

As the leakage currents are much lower than the currents of the commercial electrolysis, the working conditions of the electrodes for protection against corrosion by leakage currents may be less severe than the working conditions of the electrodes for the commercial electrolysis. However, the irregularity of the current and potential distribution over the surface of the electrodes for corrosion protection must be taken into account. This irregularity is an integral part of the work principle for the electrodes oriented along the field of the external current. The maximal current density, which can be concentrated at the end part of these electrodes, may be much higher than the average current density. Moreover, as noted in the previous chapter, the current may change in magnitude, and sometimes in its direction, which may lead to accelerated wear of the electrodes. Therefore, the final conclusions on the stability of a given electrode can be made only after its long time testing in real industrial conditions.

The problem of the selection of materials for stable cathodes is usually simpler than the selection of materials for the anodes of electrolyzers because the cathodic current in most cases provides protection to, and not destruction of, the metals. Nevertheless, such a problem exists. In particular, cathodes which are used in the processes of commercial electrolysis may undergo destruction during outage periods, when they are in contact with the aggressive environment, in the absence of protection by the cathodic current. Therefore, the material of the cathodes must be corrosion resistant in the aggressive environment where it is designed to work in the absence of the cathodic current. This is the current requirement for the cathodes designed for the protection of metals against attack by external cathodic currents.

As mentioned earlier, ceramic electrodes possess a high stability in many aggressive media under the action of both anodic and cathodic currents. However, their application as cathodes is limited due to their high cost. Researches for more available materials for stable cathodes are still in progress. For example, a nickel-molybdenum coating obtained by the thermal decomposition of alcoholic solutions of the chlorides and nitrates of these metals was proposed for cathodes of membrane electrolyzers of chlor-alkali plants [2]. This coating, applied on sheets of stainless steel, maintained its stability in 30% NaOH solution at a temperature of 178°C and current density of $4\,kA/cm^2$ over 3000 h. At a cathodic current density of $3\,kA/cm^2$, the overvoltage was only 80–100 mV. It can be assumed that cathodes with a similar coating can be used for protection against corrosion by external cathodic currents in alkaline media.

When corrosion by cathodic current occurs, it is usually significantly less intensive than corrosion by anodic currents. Therefore, it is expedient in some cases to select low cost and available materials with limited stability for manufacturing cathodes that protect metallic structures from corrosion. Periodical replacement of these cathodes may be more economical than the application of cathodes made of expensive materials.

The shape of the electrodes designed for electrocorrosion protection depends on the structural features of the object to be protected. For the protection of internal surfaces of piping and tube-shaped structures, it is expedient to use tube-shaped electrodes provided with connecting flanges. When such anodes are used, an electrochemically active coating is applied on the internal surface of the anodes and on the surface of the connecting flanges contacting with the sealing gaskets. It is important to provide a reliable electric contact between the anode and the pro-tected structure and to exclude the penetration of electrolyte to this contact.

When relatively small metallic structures are protected, the electrochemically active coating can be applied directly to the areas that are attacked by the external currents. In this case, the coated areas of the structure play the part of the electrodes and protect the whole structure in the field of the external current.

11.2. APPLICATION OF ANODES OF COMMERCIAL ELECTROLYSIS PROCESSES FOR PROTECTION AGAINST ELECTROCORROSION

The protection of the metallic equipment of technological lines against attack by external anodic currents with the help of anodes that are used in commercial electro-lysis is most available. For example, using titanium anodes with an electrochemically active metal-oxide coating in the brine lines of diaphragm electrolysis in chlor-alkali plants for this purpose is most expedient. There are detailed data on the electroche-mical characteristics of these materials in chloride media [1, 3–5] which can be used to design this protection. Nevertheless, final conclusions on the stability and efficiency of these anodes must be based on the results of industrial tests in which the working features of the anodes, as means for corrosion protection, are taken into account.

11.2.1. Neutral and acid chloride-containing media

Studies were carried out to estimate the applicability of different electrochemically active materials and coatings for protection against electrocorrosion in different technological lines of chlor–alkali electrolysis plants. As the chloride concentration in these lines varies in a wide range, the studies were carried out in NaCl solutions of different concentrations.

It was noted above, that anodes of graphite and titanium with a metal-oxide coating are characterized by a low overvoltage of chlorine evolution and by a high overvoltage of oxygen evolution. Therefore, their electrochemical activity diminishes when the chloride concentration is reduced. Moreover, under the conditions of reduced chloride concentration and elevated oxygen evolution, the corrosion stability of both the graphite and the metal-oxide coating decreases. However, these restrictions are not considered as relevant for the application of anodes of these materials for protection from corrosion by leakage currents, since these currents are very small compared with the currents of the commercial electrolysis.

It should be possible to use graphite anodes successfully for protection from electrocorrosion in both concentrated and dilute chloride solutions owing to their low cost and workability. However, their use is restricted due to the low strength and porosity of graphite. The electrolyte penetrates through the pores of tube-shaped graphite anodes. Thickening should make graphite more promising for use as anode material, since it not only reduces the porosity but also increases the strength of the graphite. An important advantage of graphite, as compared with the metal-oxide anodes, is its stability under the action of both anodic and cathodic external currents.

In this connection, studies of modified graphite MG, thickened with pyrolytic carbon to reduce its porosity [6], were carried out.

The porosity of the graphite before thickening was 25% at a statistical average pore radius of 4–5 microns and a specific surface area of $0.6 \, m^2/g$. The degree of thickening by the pyrolytic carbon was 10–12%. The volume of the pores and the specific surface area decreased after thickening by 25–30%.

The specimens for testing had the form of cylinders of diameter 10mm and length 20 mm. The electrochemical investigations were carried out in 1 g/l NaCl solution at a temperature of 90°C, in accordance with the procedure that was used in the investigations of metallic specimens.

The anodic behavior of non-thickened and thickened graphite is noticeably different (Figure 11.1, curves 1 and 2). The current density on the non-thickened graphite at potential 1.2 V is by an order of magnitude higher than on the thickened graphite. However, due to the lower slope on the polarization curve obtained on the thickened graphite (0.25 V), near potential 1.9 V, the curves converge. The differences in the anodic behavior of these materials can be explained by the difference in their porosity; the influence of the porosity decreases with the increasing potential [7]. Another possible cause of this difference is the formation of a layer of pyrolytic carbon on the graphite surface during its thickening.

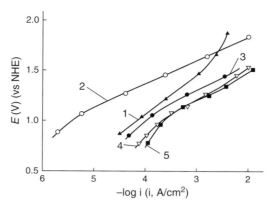

Figure 11.1   Anodic polarization plots in 1 g/l NaCl solution on graphite MG non-thickened (1) and thickened by pyrolytic carbon (2–5), without pretreatment (1, 2) and after pretreatment in 200 g/l NaOH solution at current density 10 mA/cm² during (h): 3–1; 4–6; 5–12. Temperature, 90°C.

To estimate of the surface state influence of graphite on its anodic behavior, pretreatment of the thickened graphite specimens was carried out by anodic polarization in 200 g/l NaOH solution at room temperature at a current density of 10 mA/cm² for 1, 6 and 12 h (curves 3–5). The rise of the anodic current after pretreatment over 1 and 6 h is, most probably, related to the increase in the true surface area and the removal of the pyrolytic carbon film. Further pretreatment did not have an influence the anodic behavior of the graphite.

Thus, thickening does not reduce the electrochemical activity of graphite at high anodic potentials, and the pretreatment leads to a further increase of its activity. These results encourage the consideration of thickened graphite as a promising material for the protection of metallic structures against corrosion by external currents, especially in chloride–containing aggressive media.

Besides graphite, coatings based on the oxides of manganese and cobalt have a low cost and high electrochemical activity in chloride containing environments [1, 8, 9]. However, anodes with these coatings did not find many commercial applications due to a barrier film forming under the coating during anodic polarization. According to the data [10], this effect is connected to oxygen transfer from the coating oxide to the surface of the titanium base. This leads to a barrier film forming on the titanium surface and a shift in the potential of titanium to the positive side, up to the value of its activation potential.

Prevention of the barrier film formation may be achieved by forming a stabilizing interstitial layer (underlayer) between the electrochemically active coating and the titanium base. Metals of the platinum group are most often used for the underlayer [1, 8], but their use significantly increases the cost of the anodes.

With a view to reducing the cost of the anodes, investigations were carried out on the efficiency of a stabilizing underlayer containing low cost and available materials – a mix of molybdenum and iron oxides. The underlayer was formed by thermal decomposition of a solution containing para-ammonium molybdate $(NH)_4Mo_7O_{24}\cdot4H_2O$, ferric chloride and hydrochloric acid in a muffle furnace at a temperature of 500–550°C for 20 min. The electrochemically active coating based on MnO_2 with the addition of about 5 molar % of Co_3O_4, was also formed by the thermal decomposition of a solution containing a mixture of manganese and cobalt nitrates, at temperatures of 260–280°C over 15–20 min. Procedures for preparation of the titanium surface before forming both the underlayer and electrochemically active coating, and the compositions of the solutions forming the mentioned layers are given in Ref. [11].

The studies of the anodes with the coatings of the oxides of manganese-cobalt (OMCA) have shown a low dependence of their electrochemical activity on the coating thickness. However, the life time of the coating depends on the quantity of the applied coating layers. The coating containing 2 underlayers and 10 layers of the OMCA was optimal. The optimal total thickness of such a coating was $15 \pm 2.5\,\mu m$. The life time of the coating went down at a lower thickness. On the other hand, cracking and scaling of the coating was possible when thicker coatings were used, due to internal stresses related to the differences in the coefficients of linear expansion of the titanium base and the coating.

The coating quality depends on the regime of the thermal treatment. After applying all the layers, the obtained anode has to be held for 30 min at a temperature of 380°C, to obtain complete crystallization of the coating.

It is seen from Figure 11.2 that the anodic electrochemical activity of the coating strongly depends on the concentration of the NaCl solutions and it diminishes with a decrease in this concentration. Nevertheless, in 1 g/l NaCl solution, this coating is still quite active and not less than graphite MG. A current density of 20 mA/cm^2 was reached in 300 g/l NaCl solution at potential 1.4 V and in 1 g/l NaCl solution – at potential 1.8 V.

A 260 h long test in 1 g/l NaCl solution under galvanostatic polarization by a current density of 20 mA/cm^2 has shown that no tendency to a shift in the potential to the

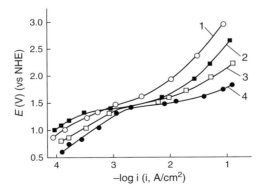

Figure 11.2 Anodic polarization plots on anode of titanium with oxide manganese-cobalt coating (OMCA) on underlayer of molybdenum and iron oxides in NaCl solutions at concentrations (g/l): 1–1; 2–10; 3–100; 4–300. Temperature, 90°C.

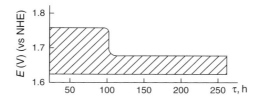

Figure 11.3 Potential chronograms of anodes of titanium with manganese–cobalt oxide coating (OMCA) on underlayer of molybdenum and iron oxides in 1 g/l NaCl solution at current density 20 mA/cm². Temperature, 90°C.

positive side of the specimens with the coating under consideration was observed, despite the preferential oxygen evolution on the surface of the tested specimens (Figure 11.3). In contrast, after 100 h of polarization, the top boundary of the oscillations in the potential shifted to the negative side, from 1.76 to 1.62 V. These results indicate that the underlayer prevents the formation of the barrier layer on the titanium surface. The majority of the weight loss of the anode took place during the first 10 h of testing. During this time the solution acquired a slightly violet color which indicated the accumulation of MnO_4^- ions. Further coloring of the solution was not observed.

The obtained results indicate the possibility of applying OMCA anodes with a stabilizing underlayer for the protection of metallic structures from corrosion by external anodic currents in chloride containing media.

11.2.2.   Chloride-sulfate media

In solutions containing sulfate ions alongside the chloride ions, metal–oxide anodes and, in particular, anodes with oxide coatings of base metals, are also used [12], despite their elevated wear.

The presence of sulfate ions in such mixed solutions leads to a shift in the potential of anodes with a coating of the oxide of ruthenium-titanium (ORTA) to the positive side. Nevertheless, ORTA anodes possess a quite high stability and electrochemical activity in acid chloride-sulfate solutions [13]. The current efficiency of chlorine evolution rises when the anodic current density on these anodes increases.

Investigations of ORTA and OMCA anodes were carried out in the nickel electrorefining solution. It is seen from Figure 11.4 that the electrochemical activity of ORTA (curve 1) in this solution is higher than that of OMCA (curve 2). The slope of the polarization curves in the Taffel area was equal, respectively, to 85 and 125 mV. A current density of 100 mA/cm² was reached on ORTA at potential 1.45 V, while on OMCA it was an order of magnitude lower. Nevertheless, the electrochemical activity of OMCA was high enough to provide corrosion protection to titanium, due to the high activation potential value of this metal.

Under galvanostatic polarization by a current density of 50 mA/cm² the potential was stabilized on ORTA at values 1.75–1.8 V and the weight loss of the anode over 25 h was insignificant. At this current density, the same initial potential value was observed on OMCA, but over the first 25 h the average corrosion rate was equal to 0.9 g/m²h. Over the next 5 h, the potential shifted to the values 2–2.2 V and the

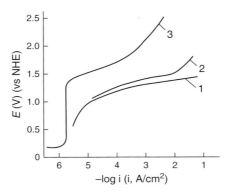

Figure 11.4 Anodic polarization plots on anodes coated by oxide ruthenium–titanium (ORTA) (1) and oxide manganese-cobalt (2) and on uncoated titanium (3) in catholyte of nickel electrorefining. Temperature, 85°C.

corrosion rate attained $2–2.5 \, \mathrm{g/m^2 \, h}$, which indicated destruction of the anode. As the anodic current density was reduced to $20 \, \mathrm{mA/cm^2}$, the potential was equal to $1.6–1.65 \, \mathrm{V}$, but the corrosion rate was still rather high – $0.25–0.30 \, \mathrm{g/m^2 \, h}$. Therefore, OMCA anodes may find only a limited application, and only at low current densities in the considered solution, in spite of a high chloride concentration. It is preferable to use ORTA anodes in chloride-sulfate solutions of nickel electrorefining.

11.3. ANODES FOR MEDIA CONTAINING SULFURIC ACID

Solutions containing sulfuric acid belong to the main media in electrochemical processes of electrorefining and electrowinning of copper, nickel, cadmium, zinc and other metals in hydrometallurgical works. Lead and its alloys are the most often applied as a material for the anodes in these processes, which have been studied in detail [14, 15] owing to the wide use of lead in lead–acid accumulators. It was found that lead dissolution and the accumulation of Pb^{2+} ions close to the anode surface takes place at the initial stage of anodic polarization. Therewith, an insoluble and non-conductive layer of lead sulfate deposits on the anode surface which brings about a rise in the potential and current density at the areas that are free of this deposit. At potential $0.8 \, \mathrm{V}$, transfer of Pb^{4+} ions becomes possible, with the formation of a salt, $Pb(SO_4)_2$. As a result of its hydrolysis, lead dioxide forms on the free areas of the anode:

$$Pb(SO_4)_2 + 2H_2O = PbO_2 + 2H_2SO_4$$

Above potential value $1.68 \, \mathrm{V}$, the standard potential of the reaction:

$$PbO_2 + 4H^+ + SO_4^{2-} + 2e^- = Pb(SO)_4 + 2H_2O$$

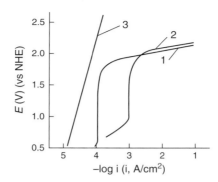

Figure 11.5   Anodic polarization plots of lead (1), platinum (2) and titanium (3) in electrolyte of copper electrorefining. Temperature, 60°C.

all the surface of the anode becomes coated with a conductive layer of PbO_2. This layer has a high electrochemical activity. Anodic processes of oxygen evolution, oxidation of SO_4^{2-} to $S_2O_8^{2-}$, as well as of hydrogen peroxide and ozone formation, pass over the surface of this layer at a high velocity. At the same time, there are data [15] that the dissolution rate of the lead increases noticeably in dilute sulfuric acid solutions.

In this connection, electrochemical investigations were carried out on lead and, for comparison, of platinum in the copper electrorefining electrolyte.

It is seen from Figure 11.5 that in the range of potentials 0.5–1.7 V the current density on lead does not depend on the potential (curve 1), due to its "salt passivity". At more positive potentials, the electrochemical activity of lead proliferates. The Taffel dependence takes place above potential 2 V, which characterizes the electrochemical processes of anodic oxidation on the surface of the formed PbO_2 layer. In this area of potentials the dependences of lead and of platinum are close to one another.

Upon galvanostatic polarization by a current density of $50\,mA/cm^2$, potentials 2.4 V and 2.2–2.3 V are stabilized, respectively, on platinum and on lead. The slightly more negative potential values of lead may be explained by the more developed surface of the PbO_2 oxide and by its anodic dissolution, which is quite high in the considered conditions (Table 11.1).

Table 11.1   Corrosion rate $(g/m^2\,h)$ of lead specimen at anodic polarization by current density $50\,mA/cm^2$

Test conditions	No. of tests					
	1	2	3	4	5	6
Without stirring	4.9	1.2	14.5	1.4	28.4	4.8
With stirring	4.0	10.7	5.4	–	–	–

The results of 5 h long tests given in Table 11.1 were obtained on a single specimen. It is seen that great fluctuations in the corrosion rate are observed during repeated tests of the specimen. This can be explained by the scaling of the formed PbO_2 layer, which has a low mechanical strength and low adhesion to the metal. A part of the current is spent on the renewed surface of lead oxidation. The fluctuations of the corrosion rate under stirring conditions decrease, since in this case the superficial oxide layer is removed more regularly.

In order to define the area in which the lead corrosion rate increases, the specimens were held at potentials 1.4, 1.7 and 2.2 V. It was found that at potential 1.4, which relates to PbO formation [16], and at 1.7 V, which accords with the inflection potential of curve 1 in Figure 11.5, an even weight increase of 1–2 mg is observed after 5 h testing of lead oxidation. Intensive corrosion occurs at potentials that are more positive than the potential of PbO_2 formation (1.8 V). The corrosion rate reached $3.6 \, g/m^2 \, h$ after 5 h of polarization at potential 2.2 V.

The results testing compounded titanium – lead and titanium – platinum specimens with surface area ratios of 1:1 and 1:2 indicate the possibility of the corrosion protection of titanium on the action of external anodic currents under the considered conditions. At an anodic current density of $50 \, mA/cm^2$, relating to the total surface area of the Ti-Pb and Ti-Pt specimens, the potentials were stabilized at values, respectively, of 2.5–2.6 and 2.6 V. The Ti and Pt weight losses were insignificant, but on Pb, the corrosion rate reached $7.2 \, g/m^2 \, h$. As the anodic current density on lead at this potential was more than three orders of magnitude higher than on titanium, practically all the current was concentrated on the lead part of the specimen, and the current density on this part was equal to $100 \, mA/cm^2$.

Thus, while providing complete protection of the titanium from corrosion by an external anodic current, the lead anode, itself, undergoes quite intensive corrosion. Therefore, anodes must have a considerable mass, in order to acheive the necessary lifetime. In addition, it is necessary to apply means for reducing the sizes of the external currents attacking the metallic structures protected by lead anodes.

11.4. DIMENSIONALLY STABLE ANODES WITH A COATING OBTAINED ON THE BASIS OF INTERMETALLIC COMPOUNDS PRODUCED IN AN ELECTRIC SPARK

11.4.1. Preconditions for the choice of materials and methods of producing the anodes: Test conditions

It was already noted that for protection from corrosion by external currents, the most promising are anodes that have a low overvoltage of oxygen evolution and maintain their stability under these conditions. When the chloride concentration in technological solutions diminishes, the electrochemical activity and stability of the above considered anodes goes down, since these anodes are preferably designated for the electrolysis of concentrated chloride solutions, and have a high overvoltage of oxygen evolution.

Magnetite is one of the anode materials that is characterized by a relatively low overvoltage of oxygen evolution and a high stability under the conditions of chlorine evolution in chloride-containing media [17]. This is the reason that magnetite anodes are used for producing chlorates. However, magnetite has an elevated electrical resistance and fragility, which limits the possibilities of its application.

Nickel possesses a low overvoltage and high stability in the process of oxygen evolution and, therefore it widely used as an anode in the processes of water electrolysis. However, nickel undergoes pitting corrosion in the presence of chlorides [18].

As noted above, the best characteristic for the anode material designated for protection from electrocorrosion is a combination of a low overvoltage of oxygen evolution and a high stability in chloride-containing solutions.

The data of Chapter 6 on the reduction of the overvoltage of oxygen evolution on a titanium alloy containing 2% nickel which is connected with the presence of the intermetallic compound Ti_2Ni in this alloy, points to the possibility of obtaining an anodic material with such characteristics. The best results were obtained with this alloy in alkaline media where it behaves as an effective anode. The presence of chloride does not cause damage to this alloy because the oxygen evolution potential is much more negative than its activation potential E_a. Example 3 of Paragraph 8.3 shows that the Ti-2% Ni alloy can be used as an effective anode for the protection of titanium structures from corrosion by an external anodic current in chlor-alkali media.

It was assumed, based on these data, that increasing the intermetallic Ti_2Ni content in the superficial metal layer would increase its electrochemical activity. This supposition was supported by the information that pure intermetallic Ti_2Ni can be used as an anode material in neutral and weakly acid chloride solutions [19–21].

The most common coating methods by which intermetallic compounds are formed on the metal surface as a result of interaction between the base and the coating metals are as follows: treating in an electric spark [22, 23] and plasma spraying of metals [24, 25]. It is known, in particular, that intermetallic compounds of different compositions are formed when titanium is treated with nickel in an electric spark [23, 26].

Corrosion and electrochemical studies were carried out of coatings obtained under the conditions of Ti treatment by Ni (Ni/Ti coatings) in an electric spark and by plasma spraying. The coatings were manufactured under free access of air, with the help of devices generating an electric spark between either vibrating or rotating electrodes of nickel and a treated (i.e., coated) metal (titanium). In a part of the experiments, coatings obtained as a result of titanium treatment by other materials were tested as follows: Co/Ti, Fe/Ti, C/Ti and Ni-C/Ti (the titanium surface was treated in succession by nickel and by carbon), W/Ti and Pb/Ti. Several experiments were done with Ni/Ti-0.2% Pd specimens, i.e., titanium alloy containing 0.2% palladium was used as the base metal.

Quite severe regimes, corresponding to discharge energies of 5.4 and 8 J, were used for the coating in most of the experiments. For several tested specimens, milder coating regimes were used, corresponding to discharge energies 0.1, 0.24, 0.84 and 2.7 J, were used.

Specimens with 200 micron thick Ni/Ti coatings were prepared by plasma spraying in a nitrogen-argon environment.

Spade-shaped, 1–2.5 mm thick, specimens with a working surface area $\sim$8 cm^2, were tested in most of the experiments. In some of the experiments, cylindrical specimens with a working surface areas of a same size were tested.

Electrochemical studies were carried out at potentiodynamic (0.72–3.6 V/h), potentio- and galvanostatic polarization over 5–25 h, with interruptions after each 5 h overnight.

Additional experimental conditions will be detailed during the discussion on the obtained results.

11.4.2.   Investigations in chloride solutions

In the first stage of the studies, the potentials and weight losses of specimens coated in an electric spark after 5 h long testing in chlorine-saturated 1 g/l NaCl solution at a temperature of 90°C without polarization were measured (Table 11.2).

It is seen from the table that the potentials of the specimens with coatings Pb/Ti and W/Ti coatings did not change significantly during the corrosion test. A strong shift in the potential of other coated specimens to the positive side points to their self-passivation in the solution, which has a high redox potential. On the specimen with the Ni/Ti coating after 5 h exposure, the potential continued to shift to the positive side. Weight losses of the Fe/Ti, Co/Ti and Ni/Ti specimens were significant after the first 5 h long tests, but were close to zero after repeated corrosion tests. After repeated potentiodynamic polarization, the weight losses of these three types of specimens also dropped by more than one order of magnitude (Table 11.2, two last columns). The weight losses of the specimens with the Pb/Ti and W/Ti coatings remained significant after repeated corrosion tests, which indicated the active dissolution of these coatings.

Among the considered coatings, Ni/Ti possesses the highest electrochemical activity of oxygen evolution (Figure 11.6). The slope of the potentiodynamic polarization curve obtained on this coating, at a scanning rate of 0.72 V/h above potential 1.4 V, was equal to 170–180 mV.

In the first potentiodynamic polarization curves obtained on Ni/Ti and Fe/Ti coatings, loops appeared at potentials more negative than the oxygen evolution potential (Figure 11.7, curves 1 and 3). Upon repeated polarization of the specimens in the same conditions, the loops did not form. Also, in the repeat test, the slope of the curve did not change. These results show that in the coatings obtained in an electric spark that dissolve at potentials more negative than the oxygen evolution potential and components which maintain their stability at the oxygen evolution potential and which determine the anodic electrochemical activity of the Ni/Ti coating.

The polarization curve obtained on the W/Ti coating had a steep slope, due to the high ohmic resistance of the coating: even at a low polarization current, the output resistance of the potentiostat sharply increased.

Proliferation of the current on the Co/Ti coating was observed at potential 1.2 V, which is more negative than the chlorine evolution potential and more negative than the potential of current proliferation on the Ni/Ti coating. This may be related to a lower overvoltage of oxygen evolution on Co/Ti. However, with the increasing potential the polarizability of the Co/Ti coating increased and its electrochemical

Table 11.2 Electrode potentials and weight losses of specimens coated by an electric spark in chlorine saturated NaCl solution. Temperature, 90°C

No.	Coating	Electrode potential		Weight losses (mg)		
		Initial	Final	After corrosion tests	After polarization	
					First test	Repeated test
1	Pb/Ti	−0.24	−0.22	9.1	19.9	–
2	W/Ti	−0.50	−0.42	1.9	0.5	0.6
3	Fe/Ti	−0.30	0.15	1.2	10.9	0.4
4	Co/Ti	−0.10	1.04	2.2	10.9	0.2
5	Ni/Ti	−0.10	0.65	10.2	19.2	0.3

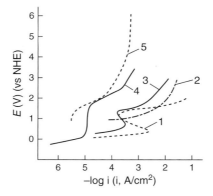

Figure 11.6 Anodic polarization plots in chlorine-saturated 1 g/l NaCl solution on titanium with coatings: 1 – Ni/Ti; 2 – Co/Ti; 3 – Fe/Ti; 4 – W/Ti; 5 – uncoated Ti. Temperature, 90°C.

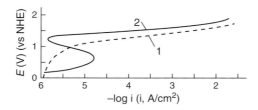

Figure 11.7 Anodic polarization plots on titanium with electric spark Ni/Ti coating in chlorine-saturated NaCl solutions with NaCl concentrations, g/l: 1–1 (repeated polarization); 2–300 (first polarization). Temperature, 90°C.

activity became lower than on Ni/Ti. On the Ni/Ti coating, no growth in the polarizability was observed up to potential 2 V. Most probably, simultaneous evolution of oxygen and chlorine took place on this coating, with a preferential oxygen evolution. This assumption was proved, as the polarization curves obtained in

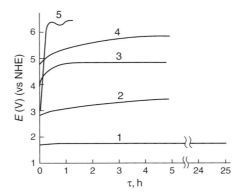

Figure 11.8 Chronograms of potential at polarization in 1 g/l NaCl solution of titanium with coatings: 1 – Ni/Ti; 2 – Co/Ti; 3 – Pb/Ti; 4 – W/Ti; 5 – uncoated titanium. Temperature, 90°C.

solutions 1 and 300 g/l are compared (Figure 11.7). At the fairly similar slopes of both of these polarization curves in the Taffel area, the curve obtained in the 300 g/l solution is disposed at potentials that are 50–60 mV more positive.

Galvanostatic tests in 1 g/l NaCl solution at a temperature of 90°C proved that the Ni/Ti coating was the most stable. At a current density of 20 mA/cm² the potential of a specimen with this coating was equal to 1.76 V and remained stable over 25 h (Figure 11.8, curve 1). The weight losses of the specimen (previously pretreated by potentiodynamic polarization) after this 25 h long test were quite small – only 0.35 mg.

After 5 h long polarization of specimens with a Co/Ti coating (curve 2), the potential reached a value of 3 V and had a tendency to further shifting to the positive side.

On the specimens with Pb/Ti and W/Ti coatings, the potential reached values of, respectively, 4.8 and 5.7 V (curves 3 and 4) and the coating was destroyed.

The data obtained on the high anodic electrochemical activity and stability of the Ni/Ti electric spark coating promoted more detailed investigations of the composition and structure of this coating. X-ray analysis of the coating was carried out before and after 10 h long anodic galvanostatic polarization at a current density of 20 mA/cm² in a 1 g/l NaCl solution, at a temperature of 90°C.

The following components were found in the coating before the polarization (Figure 11.9a): Ti, Ni, traces of the oxides TiO and NiO, TiN, and the intermetallic compounds Ti_2Ni, TiNi and $TiNi_3$. As seen from Figure 11.9a, all the components are identified by several peaks.

Identifying TiN and TiNi was hampered, because the two clearest peaks relating to the interplanar spacing 1.505 and 2.12 (respectively, angles 2θ equal to 61.4° and 42.6°) coincide for these components, and peak 2.44 (angle $2\theta = 36.8°$) which relates only to the TiN is low. Data on the formation of this compound were communicated in work [27]. The TiN formation was confirmed by X-ray analysis

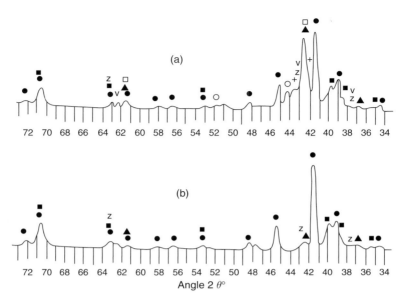

Figure 11.9 Roentgenogram of Ni/Ti coating (a) – before polarization; (b) – after polarization: ▲– TiN, ■ – Ti, • – Ti_2Ni, z – TiO, ○ – Ni, □ – TiNi, v – NiO, + – $TiNi_3$.

of the superficial layer obtained by treatment of the titanium surface in an electric spark by a titanium electrode, similarly to the treatment that was described in Ref. [28].

A sharp decrease of peak 2.12, related to TiNi and TiN, was noted after polarization (Figure 11.9b). The fact that peak 2.44, which relates to TiN, remained after the polarization, indicates that just this compound is stable under anodic polarization. This was proved by the results of testing 0.1 mm thick specimens which were cut from a TiNi foil. These specimens were completely destroyed in similar 5 h long tests.

The results given in Figure 11.9 show that Ni, NiO and intermetallic $TiNi_3$ also dissolve in the process of anodic polarization. It is most probable that the weight losses and appearance of a loop on the anodic potentiodynamic polarization curve during the first polarization, are explained by the active dissolution of all of these components. Some of them, in particular TiNi and Ni, intensively dissolve in chlorine-saturated solution, even without polarization, and this explains the weight losses which took place in the first corrosion tests of specimens with Ni/Ti coating, carried out in this solution.

The height of the peaks related to intermetallic Ti_2Ni did not diminish after the polarization. This proves that the intermetallic Ti_2Ni is stable under anodic polarization in chloride-containing solutions, under the conditions of preferential oxygen evolution. Retention of the electrochemical activity by the coating and a sharp decrease in the weight losses on repeated polarization, confirm that the electrochemical activity is determined by the presence of Ti_2Ni. The results of tests on a pure intermetallic Ti_2Ni [20, 21], which proved its high stability and electrochemical activity, are in line with this conclusion.

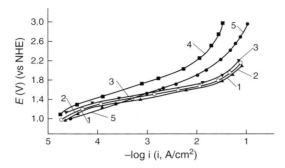

Figure 11.10   Anodic polarization plots in 1 g/l NaCl solution at temperature 90°C: 1–3 – on Ni/Ti coating, obtained in an electric spark; 4, 5 – on intermetallic Ti₂Ni. 2, 3, 5 – after 8 h of outdoor oxidation at temperatures, respectively, 500, 900 and 800°C.

Oxidation of the superficial coating layers is possible in the process of outdoor coating by an electric spark. There are data [29] that intermetallic Ti$_2$Ni actively absorbs oxygen and can form a stable phase Ti$_4$Ni$_2$O which has lattice parameters close to the lattice parameters of Ti$_2$Ni. The proximity of these parameters makes it impossible by X-ray analysis to determine whether the intermetallic Ti$_2$Ni of the coating is oxidized. In order to estimate the influence of the oxidation, the coated specimens were oxidized outdoors at temperatures of 500 and 900°C for 8 h.

The potentiodynamic (3.6 V/h) polarization curves obtained on the specimens with the Ni/Ti coating (Figure 11.10, curves 1–3), before and after oxidation, practically coincide. This result suggests that the intermetallic Ti$_2$Ni in the coating already has an oxidized form and additional oxidation does not noticeably change the surface state. In order to determine whether the oxidation of the intermetallic Ti$_2$Ni influences its electrochemical activity, potentiodynamic polarization plots were obtained on specimens of pure intermetallic Ti$_2$Ni, non-oxidized, and oxidized in the air at 800°C (curves 4 and 5). It is seen that the current density increases on the oxidized surface of the intermetallic Ti$_2$Ni increases and approximates the current density obtained on the Ni/Ti coating. These results prove that the obtained coating has an oxidized form and oxidation increases the electrochemical activity of the intermetallic Ti$_2$Ni.

The considered results on the high electrochemical activity and stability of the Ni/Ti coating obtained in an electric spark and on determining the role of the oxidized intermetallic Ti$_2$Ni in these properties, corroborate the data given in Refs. [30, 31].

Alongside, the composition analysis, the depth-wise distribution of the components in the Ni/Ti coating and some specific properties of the coating were studied.

The composition of the coating through its thickness was determined by layer-by-layer etching in 40% sulfuric acid solution at a temperature of 90°C, followed by colorimetric analysis of the etching solution for the titanium and nickel ion content. It was established (Figure 11.11) that the outer 0.05 μm thick coating layer was enriched with nickel. In the next 15–25 μm thick layer, its content was 12–20 wt.%

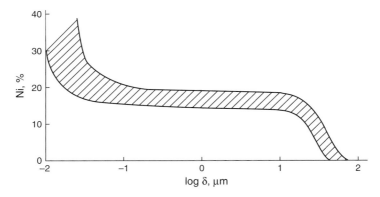

Figure 11.11 Nickel content (wt. %) through the coating thickness.

(atomic ratio Ti:Ni $= 8.9{:}4.7)^{\star}$. Further down, the nickel content rapidly decreased. The total coating thickness enriched with nickel reached 30–50 μm. The atomic ratio of Ti:Ni > 2 (i.e., higher than in the intermetallic Ti$_2$Ni) in the layer with the stable composition may be connected not only with the elevated titanium content in the coating, but also with etching of the titanium base through the pores of the coating.

With anodic polarization in 1 g/l NaCl solution at 90°C in galvanostatic (10 mA/cm^2), and potentiostatic (1.9 V), regimes there was also a preferential transfer of nickel into the solution in the initial period. Correspondingly, a rapid shift of the potential to the positive direction, up to the values of oxygen and chlorine evolution, or a decrease in the current density to a stable value of ~1 mA/cm^2, occurred. During the first 3–5 h, the Ti:Ni atomic ratio in the solution was 0.15–0.5, then it increased to 2–3, and then, after 8–10 h of polarization, it reached 6–9. In the period of preferential nickel dissolution (related, as shown above, to the dissolution of Ni, NiO, TiNi$_3$ and TiNi) there was significant, up to several milligrams, weight loss of the specimens.

It was revealed that cathodic polarization in the same solution at potential -1.0 V, leads to intense destruction of the coating. The current density was ~1 mA/cm^2 in these conditions and the weight loss in the 5 h long test was more than 10 mg. The destroyed coating passed into the solution in the form of a sludge with a particle size of 0.1–0.3 mm and with Ti:Ni atomic ratio of 2.5–4. Since, after dissolution of the 0.5 μm thick outer layer enriched with nickel, the coating consists mainly of intermetallic Ti$_2$Ni, it is evident that the sludge particles represent an alloy of Ti$_2$Ni with titanium. Nevertheless, X-ray analysis has shown that the peaks corresponding to Ti$_2$Ni and Ti are very weakly defined (Figure 11.12c), which provides a basis for

---

*Shown in the text is the range of values obtained within the limits of scatter for data which is apparently connected with a manual coating application, i.e., under conditions when it is impossible to strictly control the relative positioning of the electrodes, the rate and movement path for the electrode-tool, the pressure of the anode (Ni) to the cathode (Ti) surface, etc. Given in Figure 11.11 are the lower and upper boundaries within whose limits data were obtained for the coating composition.

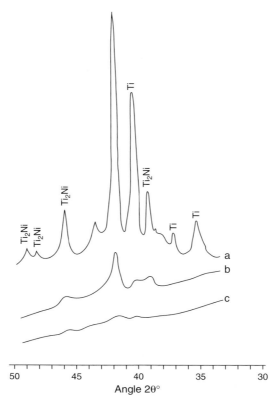

Figure 11.12 X-ray pictures of Ni/Ti coating (a) and of accumulated sludge (b, c) after cathodic polarization of coated specimen at a potential −1 V. b − sludge after annealing in helium 1 h at 1000°C; c − non-annealed sludge.

suggesting that the particles have an amorphous or fine-crystalline (quasi-amorphous) structure. In order to verify this suggestion, a recrystallization annealing of sludge particles in a helium atmosphere was carried out at 1000°C for 1 h. Under such annealing, the amorphous metal returns to a crystalline structure [32, 33]. In fact, after annealing, the Ti$_2$Ni peak increased strongly and the Ti peak increased slightly (Figure 11.12b). This may be explained by crystallization of the amorphous phase (if there is one) during heading of the specimen or by an increase in the size of the quasi-amorphous crystalline particles to those values which give this result in X-ray structural analysis.

For the coatings obtained under a mild regime of electric discharge (0.24 J), a low Ti$_2$Ni peak intensity and a high background level were most typical (Figure 11.13a). After annealing the specimens with this coating in air for 2 h at a temperature of 800°C, the intensity of the peaks sharply increased (Figure 11.13b). This indicates the presence of amorphous or quasi-amorphous structures in the coating.

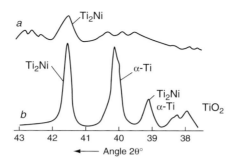

Figure 11.13 Fragment of X-ray pictures of coating obtained in a mild electric spark discharge: a – before annealing; b – after annealing in air 2 h at temperature 800°C.

The formation of an amorphous or quasi-amorphous phase may be connected with the fact that at the point of spark discharge a rapid cooling takes place, resulting in an intense outflow of heat from this point into the mass of the metal. The phase that transfers into solution in the form of a sludge evidently exhibits a higher stability than the crystalline phase which surrounds it and undergoes destruction under these conditions. It can be suggested that just the etching along the crystalline boundaries of the more stable phase leads to its detachment from the coating and conversion into a sludge. The presence of the crystalline phase in the coating is proved by the X-ray picture (Figure 11.12a): the crystalline phase was detected in the coating after cathodic polarization.

As, in accordance with the previous assumption, the intermetallic Ti_2Ni in the coating has an oxidized form Ti_4Ni_2O, it was assumed that destruction of the coating under cathodic polarization may be connected with the reduction of this oxide. However, the pure non-oxidized intermetallic compound Ti_2Ni also undergoes destruction, with the transfer of sludge particles into solution, probably also as a result of etching out of these particles along the crystalline boundaries. It is possible that the latter effect is connected with the hydrogenation of Ti_2Ni under cathodic polarization. It is well known [34] that intermetallic compounds of titanium with nickel exhibit a high capacity to absorb hydrogen.

A comparison of X-ray pictures obtained under mild (0.24 J) and severe (8 J) coating regimes shows that both the content of the intermetallic Ti_2Ni and the share of the crystalline phase rise when the discharge energy increases. This may be related to the elevated temperature of the metal during severe regimes, which leads to crystallization or to a crystal growth in the coating. The more severe the coating regime, the higher is the content of Ti_2Ni in the coating and the coating rate. The duration of metal surface treatment at the discharge energy 8 J, must be no more than 15 s/cm^2. As the treatment time is longer, the content of Ti_2Ni in the coating decreases and the shares of TiNi and (to a lesser extent) of TiNi$_3$ rises. Hence, it is expedient to increase the rate of coating under severe regimes of electric discharge, which extends the possibilities of manufacturing anodes by coating in an electric spark using high productivity installations.

The obtained data indicate the high sensitivity of some structural intermetallic components to the action of a cathodic current. Therefore, the revealed difference in the stability of such different phases as amorphous and crystalline structures is so pronounced.

The amorphisation of phases formed in an electric spark may be used as one of ways for regulating the properties of the anodes with a Ni/Ti coating. Extraction of the phases with an amorphous structure by cathodic polarization enables the study of these phases in the process of improving coatings.

11.4.3.   Investigations in chloride-alkali solutions

Potentiodynamic (0.72 V/h) anodic polarization curves obtained in a solution of 120 g/l NaOH + 200 g/l NaCl on Ni/Ti, C/Ti, C-Ni/Ti and Co/Ti coatings are rather close to one another (Figure 11.14). The peak of the current on the C/Ti coating at potential 0.2 V is probably related to the change in the coating ability to passivation. On all these coatings, proliferation of the current takes place at potentials above 0.4 V, which is connected to oxygen evolution. Above potential 0.6 V, the dependences have a Taffel's character of a low slope: for C/Ti and Ni-C/Ti it is equal to 70 mV, for Ni/Ti and Co/Ti it is equal, respectively, to 70–80 and 80–90 mV.

The potential of chlorine evolution was not reached on any of these coatings due to their high electrochemical activity at the potential of oxygen evolution.

After a first anodic polarization of all of these specimens, weight losses of 0.1–0.15 mg/cm^2 took place. As in the experiments done in neutral solutions, after a repeat polarization of the same specimens, the weight losses sharply decreased.

The potentiodynamic polarization curve on Ni/Ti in pure 120 g/l NaOH practically coincides with curve 3 (Figure 11.14) obtained in a chloride-alkali solution with the same concentration of alkali. When the alkali concentration was reduced to 4 g/l,

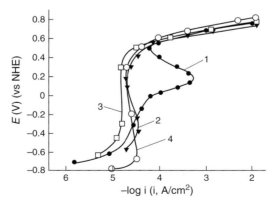

Figure   11.14   Anodic   polarization   plots   in   solution   120 g/l NaOH + 200 g/l   NaCl   at temperature 90°C on titanium with coatings obtained in an electric spark: 1 – C/Ti; 2 – Ni-C/ Ti; 3 – Ni/Ti; 4 – Co/Ti.

Table 11.3   Potentials (V) established on the coated specimens at galvanostatic polarization in solution 120 g/l NaOH + 200 g/l NaCl at temperature 90°C

Current density (mA/cm^2)	Coating					
	Ni/Ti	C/Ti	Ni–C/Ti	Co/Ti	Ni/(Ti–0.2%Pd)	Ni/Ti, plasma spraying
20	0.70–0.78	0.80–0.90	0.73–0.77	0.78–0.80	0.76–0.81	0.62–0.66
50	0.80–0.87	0.99–1.00	0.79–0.88	0.79–0.80	–	0.73–0.80
100	0.90–1.10	4.30–4.50	0.98–1.04	0.95–1.05	1.07–1.22	0.86–0.94

the oxygen evolution potential shifted to 0.7 V. Hence, the anodic behavior of the coatings in the considered conditions is determined by the alkali concentration.

The oxygen evolution potential was established on the specimens after several minutes of galvanostatic polarization (Table 11.3) and remained constant throughout the 5 h long experiment.

The shift in the potential of the specimen with the C/Ti coating to 4.5 V after 5 h long testing resulted in destruction of the coating at a current density of 100 mA/cm^2. The 25 h long test of the Ni/Ti coating at a current density of 20 mA/cm^2 and 5 h long tests at a stepwise current density increase from 20 to 50 and to 100 mA/cm^2, have shown that at all the mentioned current densities, the weight losses in the first tests were equal to 0.3–2 mg/cm^2 and went down to 0.01–0.02 mg/cm^2 in the repeat tests done on the same specimens. In a test of a specimen with a Ni/Ti coating for 10 h at 200 mA/cm^2, after its polarization at 50 and 100 mA/cm^2, a potential of 1.1 V was established and the weight loss was equal to 0.1 mg/cm^2. The weight loss after the second 5 h was only 0.01 mg/cm^2.

When an alloy of Ti with the addition of 0.2% Pd was used as the base metal coated in an electric spark by nickel, i.e., on a coating of Ni/(Ti + 0.2% Pd), the potential was slightly more positive than on Ni/Ti. At a current density of 100 mA/cm^2, its weight loss was very small, only 0.01 mg/cm^2 over 5 h. At a current density of 1000 mA/cm^2, the potential shifted to 1.8–2.2 V and the weight loss over 10 h rose to 0.2 mg/cm^2.

The Ni/Ti coating obtained by plasma spraying corroded under galvanostatic polarization at a significant rate: 0.2 and 0.7 mg/cm^2 at current densities, respectively, of 10 and 20 mA/cm^2. It is probable that the major part of the nickel did not form an intermetallic compound with the titanium base, or formed unstable compounds. The relatively low potential values established on this coating may be explained by the dissolution of nickel or to unstable nickel compounds (Table 11.3).

As seen from Figure 11.11, the content of nickel in the coating layer at a depth of more than 30 microns is no less than 2%. The results of the studies on a titanium alloy containing 2% nickel given in Section 6.2, show that 0.2% nickel is enough to provide a high electrochemical activity of the coating in chloride–alkali media.

Thus, the results of galvanostatic studies, considered in combination with these data, show that anodes with a Ni/Ti coating obtained in an electric spark can provide long-time effective protection of titanium against corrosion by external anodic currents in a chloride–alkali environment.

Detailed information on the results of investigations of the conditions of formation, structural features, corrosion and electrochemical behavior of the coatings obtained in an electric spark and containing intermetallic compounds are given in works [31–33, 35–39]. The fact that the Ti_2Ni compound formed in the electric spark is the major component of the stable and electrochemically active coating, served as the basis for the development of a method of protection from corrosion by external anodic currents in chloride, chloride–alkali and alkaline media [40].

11.5. STABLE CATHODES

11.5.1. Cathodes for media containing dissolved chlorine and for chloride media

It was shown in Chapter 7 that the accumulation of oxygen–chlorine compounds (hypochlorites and chlorates) in solutions containing dissolved chlorine, hinders the hydrogenation and corrosion of titanium. This result was used as a precondition for development of stable titanium cathodes for media containing dissolved chlorine. The idea of installing such cathodes was based on the fact that in stagnant zones (zones having a limited exchange with the bulk of the solution), which are artificially produced near the surface areas of the titanium structure attacked by external cathodic currents, the accumulation of oxygen–chlorine compounds takes place. These areas play the role of stable cathodes.

The idea was tested under laboratory conditions on a titanium specimen pre-radiated in a nuclear reactor. Spade-shaped specimens with a working surface area of 3 to 4 cm^2, cut from 0.15 mm thick foil, were prepared as described in Refs. [41, 42]. The experiments were performed in flowing and stirred chlorine-saturated, 260 g/l NaCl solution (chloranolyte) at a temperature of 90°C (the electrolysis with mercury cathode in chlor-alkali electrolysis plants).

The tested specimen was enclosed in a thin, rectangular Teflon case (specimen holder), as shown in Figure 11.15. The length and height of the slot inside the holder was in accordance with the dimensions of the specimen and its width was 2 mm so that the gap between the specimen and the inner walls of the holder was about 1 mm on either side of the specimen. Each of these side walls was perforated with 15 holes of 0.5 mm diameter, to provide a slow exchange between the solutions inside and outside the holder. The end of a Luggin capillary of a reference electrode was introduced inside one of the holes and a glass capillary of external diameter 0.5 mm, for taking samples from the slot, was introduced into another.

The tests were carried out at a constant imposed potential equal to -0.4 V, at which, according to the data of Chapter 7, the corrosion rate of titanium under the considered conditions is the maximal. The rates of corrosion and hydrogenation were measured by a radiochemical method.

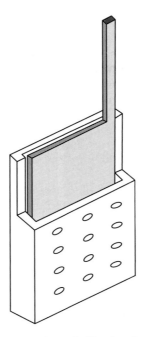

Figure 11.15 Perforated Teflon case specimen holder (explanations are given in the text) (see color plate 9).

It was found that the corrosion rate of the specimen inside the holder was low and did not change during the test (Figure 11.16). Hydrogenation was also low, only $7.3 \cdot 10^{-8}$ g-mol/cm^2. The concentrations of hypochlorite and chlorate inside the slot after 1 h of polarization reached, respectively, 0.31 and 0.42 g/l.

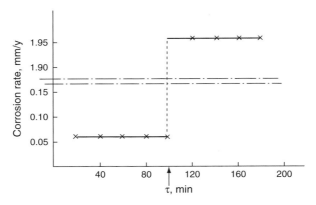

Figure 11.16 Corrosion rate vs polarization time τ. Arrow indicates the moment when the case was removed from the specimen.

On removal of the case, the corrosion rate of the titanium immediately increased by about 30 times, to the values of the corrosion rate in the bulk of a flowing and stirred solution (Figure 11.16), and the hydrogenation of the specimen also increased by more that one order of magnitude.

These results unequivocally point to the possibility of implementing stable titanium cathodes for chlorine saturated solutions by artificially forming stagnant zones near the surface of the protected metal.

As constructions in the form of tubes are mainly attacked by the external currents, a stable titanium cathode was developed to protect the internal surface of titanium piping end areas (Figure 11.17a) attacked by an external cathodic current [43].

A hollow cylindrical perforated shell, 1, made of a corrosion-resistant insulating material (e.g., Teflon) is inserted inside the terminal of a titanium pipe, at the area of attack by an external cathodic current. An annular gap, inside of which oxygen-chlorine compounds accumulate, is formed between the internal surface of the pipe wall and the external surface of the shell. A slow solution exchange between the annular space and the volume inside the pipe takes place through the perforations, 2, and through the entrance into the annular space. The terminal pipe area accepts the major part of the external cathodic current. Nevertheless, it maintains its corrosion stability under the attack of the external cathodic current I_c, due to the accumulation of oxygen-chlorine compounds inside the annular space. Thus, it plays the role of a stable cathode which protects the whole titanium pipe from corrosion. The drop in the potential from the maximal negative value (which may be more negative than the potential of titanium cathodic activation E_{ca}) at the pipe end, to some safe potential value, takes place along the area of the annular gap.

For chloride-containing solutions which do not contain dissolved chlorine, a stable cathode with a stagnant zone was also developed (Figure 11.17b). It includes a hollow shell, 1, identical to the above described one. Two double-flanged sleeves, 3 and 4, made of titanium or other conductive material which is corrosion

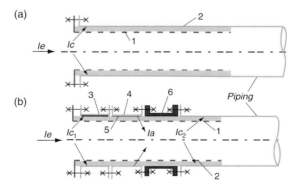

Figure 11.17 Design of a stable titanium cathode with a stagnant zone for protection of titanium piping from corrosion by cathodic external current in media containing: (a) – dissolved chlorine and chlor ions; (b) – chlor ions only (explanations are given in the text) (see color plate 10).

resistant in the medium, are installed at the pipe end. An electrochemically active and corrosion-proof coating, 5, possessing a low overvoltage of chlorine evolution (e.g. platinum, platinum group metals, oxide ruthenium-titanium) are applied to the inner surface of sleeve 4. Sleeves 3 and 4 are electrically interconnected via connecting bolts but insulated from the pipe end by insulating insert 6. Instead of the two sleeves 3 and 4, one combined sleeve may be used. In this case a part of the inner surface has the above-mentioned coating. First the external cathodic current I_e enters into sleeve 3, i.e. $I_e \approx I_{C1}$. The current passes to sleeve 4 and leaks off from coating 5 into the electrolyte as anodic current I_a. This current is spent on the oxidation of chlor-ions to molecular chlorine. The cathodic current I_{C2} flows from the solution of the annular gap into the terminal pipe area, which plays the role of a second stable cathode. Ions OH^- form on the areas of action of the cathodic currents Ic_1 and Ic_2. These ions react with the molecular chlorine generated on the anodically active coating 5, and form ions of hypochlorite and chlorate, which protect sleeve 3 and the terminal pipe area. This protected area can be considered as a stable cathode. The potential rises along this terminal pipe area disposed in the annular gap. Neither hydrogenation nor corrosion of titanium occurs beyond this gap, where a safe potential value of the titanium is attained.

11.5.2. Titanium cathodes for other aggressive media

The considered stable titanium cathodes are applicable for media containing dissolved chlorine; for chloride solutions they can be used only in a complicated form. Although these media are wide spread, the possibilities of the application of such complicated cathodes are relatively limited. Therefore, it was of interest to develop more universal stable titanium cathodes for different aggressive media.

There are data that, along with oxygen–chlorine compounds, other oxidizers are also able to inhibit titanium hydrogenation [44]. It is shown in Ref. [45] that the addition of 1 g/l of potassium dichromate to a $K_2CO_3 + KOH$ solution reduces titanium hydrogenation by about one order of magnitude. The authors explain this effect by the interaction of dichromate ions with adsorbed hydrogen atoms that are formed on the metal surface in the process of discharging hydrogen ions. This leads to a reduction in the flow of the hydrogen atoms into the metal.

Oxygen–chlorine compounds, which were used in the above-mentioned protection principle, are not introduced into the solution, but are generated by an external current and accumulate close to the protected areas of the metallic structure. The introduction of an oxidizer into the solution as a corrosion inhibitor [45] can find only limited application since it contaminates the solution and makes the protection more expensive.

The fact that oxidizers inhibit the hydrogenation and corrosion of titanium under cathodic polarization can be considered as a precondition in the search for other ways of applying them for corrosion protection. However, first it was necessary to elucidate the causes and the mechanism of this effect. For this purpose, investigations of titanium hydrogenation were carried out in a solution of 260 g/l

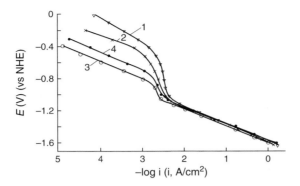

Figure 11.18   Cathodic polarization plots on titanium in 260 g/l NaCl solution (pH 2) in the presence of anions−oxidizers: 1 − ClO⁻; 2 − ClO₃⁻; 3 − Cr₂O₇²⁻; 4 − without oxidizer. Temperature 90°C.

NaCl at pH 2 and a temperature of 90°C, in the presence of the oxidizers, sodium hypochlorite, sodium chlorate and sodium dichromate, at a concentration of $5 \cdot 10^{-3}$ mol.

The share of the current that was spent for the reduction of the oxidizers was assessed from cathodic polarization curves (Figure 11.18), which can be divided into three areas. In the potential range (0.0 to −0.75) V, the kinetic area of hydrogen ion reduction, the highest magnitudes of current density (as compared with the basic solution, represented by curve 4) take place in the presence of ClO^- ions (curve 1). In a solution containing ClO_3^- ions, the current densities are lower (curve 2), but still remain an order of magnitude higher than in the basic solution. As to the solution containing $Cr_2O_7^{2-}$ (curve 3), the current densities are even lower than in the basic solution. In the areas of the limiting diffusion current of hydrogen reduction (−0.75 to −1.1) V, these dependences are much less obvious. In the area of water decomposition, all the curves practically coincide, i.e., the share of the current for hypochlorite and chlorate reduction is insignificant at these potentials.

Studies at potentials −0.4, −1.0 and −1.4 V, which correspond to each of the three mentioned areas on the cathodic polarization curves given in Figure 11.18, have shown that all the considered oxidizers inhibit titanium hydrogenation (Table 11.4), but with an increasing potential, the degree of inhibition diminishes.

Table 11.4   Hydrogenation (g-mol/cm²) of titanium in 260 g/l NaCl at pH 2 and temperature 90°C in the presence of oxidizers (5×10^{-3} mol)

Potential	Oxidizer			
	Without oxidizer	NaClO	NaClO₃	Na₂Cr₂O₇
−0.4	4.3×10^{-8}	8.9×10^{-10}	9.1×10^{-10}	1.2×10^{-9}
−1.0	5.6×10^{-7}	3.7×10^{-8}	6.4×10^{-8}	1.9×10^{-8}
−1.4	3.8×10^{-6}	5.2×10^{-7}	2.9×10^{-6}	9.5×10^{-7}

Two groups of titanium specimens were cathodically polarized in 260 g/l NaCl solution at potential -0.4 V, over 3 h: with and without preliminary cathodic polarization in solutions containing 5×10^{-3} mol NaClO and 5×10^{-3} mol NaClO$_3$. It was found that the hydrogenation values of both these groups of specimens were close to one another. This result proves that an oxide film which could hinder the hydrogenation does not form on the metal surface in the presence of hypochlorite and chlorate ions. The formation of such a film at more negative potentials is still less probable [46]. The effect of hydrogenation inhibition in the conditions under consideration may be explained by the oxidation of the hydrogen ions, which are becoming adsorbed on the metal surface, by hypochlorite and chlorate ions. It was noted in Ref. [47] that oxidation of hydrogen atoms decreases their flow into the metal.

In Auger-spectrometric investigations of the surface of the titanium specimens after cathodic polarization in solutions containing hypochlorite and chlorate, no compounds containing chlorine or sodium, which should be able to inhibit the hydrogenation process, were found. At the same time, the Auger-spectrograms carried out after cathodic polarization in the presence of dichromate-ions, detected chromium on the titanium surface. This may indicate the formation of a chromium-containing film which could serve as a barrier, hindering hydrogen penetration into the metal. In order to verify this assumption, a study of titanium hydrogenation was carried out after cathodic polarization in a 260 g/l NaCl solution at pH 2 and a temperature of 90°C, after previous cathodic deposition of a chromium compound on its surface. The said cathodic deposition was carried out at potential -1 V, in the same solution to which 5×10^{-3} mol of the radioactive salt Na$_2^{(51)}$Cr$_2$O$_7$ was added. The adsorption value was measured by the radioactive isotope $^{(51)}$Cr (half-life 27.8 days), based on the proportionality of the counting rate of the specimens (impulse/sec) to the quantity of absorbed dichromate-ions.

It was found that the adsorption of chromium-containing compounds took place in a wide range of potentials, and also in the absence of the external cathodic current, at corrosion potential E_s (Figure 11.19). The maximum amount of absorption occurred at

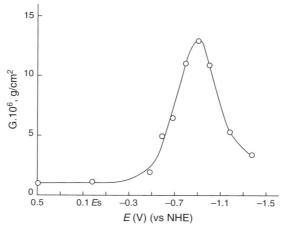

Figure 11.19   Dependence of chromium adsorption (G) on potential in 260 g/l NaCl solution (pH 2) containing 5×10^{-3} mol Na$_2$Cr$_2$O$_7$. Temperature, 90°C.

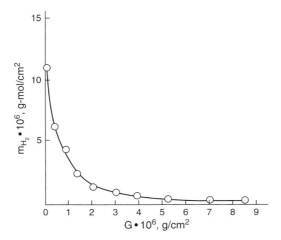

Figure 11.20 Dependence of hydrogenation m_{H2} on amount of chromium adsorption (G) in 260 g/l NaCl solution (pH 2) on titanium specimen polarized at potential -1.4 V during 5 h. Temperature 90°C.

potential -1.0 V. The highest absorption rate at this potential was observed during the first 10 min of the experiment, and then it decreased and stabilized after 1.5–2 h. A yellow deposit was formed on the metal surface after the experiment which could be easily removed by rubbing with filter paper. After such a rubbing, the content of chromium in the superficial layer was reduced by about one half. It can be assumed that initially the dichromate ions are reduced on the metal surface and form a chemisorbed film of trivalent chromium. The external, easily removable part of this film could be formed as a result of the physical adsorption of the dichromate ion [47].

As the adsorption increased to an amount of 5×10^{-6} g/cm^2, the hydrogenation fell by almost two orders of magnitude (Figure 11.20), and remained constant at higher adsorption amounts. It is probable that the efficiency of the formed superficial layer, as a barrier to hydrogenation, is determined by the chemisorbed film that is formed in the initial period of cathodic polarization.

The protective film, which hinders hydrogenation and, consequently, corrosion of titanium under cathodic polarization, is formed on its surface in a short period of time, about 10 min. This film provides titanium protection either in media containing dissolved chlorine or in media of other compositions. The use of stable cathodes obtained by the treatment of titanium in chromate-containing solutions extends the possibilities of protecting titanium structures.

11.5.3. Possibilities of increasing the stability of titanium cathode blanks

Data on hydrogenation inhibition by the treatment of titanium in a dichromate-containing solution promoted investigations on the possibility of using this treatment for the protection of titanium cathode blanks from hydrogenation. The problem of their hydrogenation in chloride-sulfate solution, under the conditions of nickel electrorefining, was considered in Chapter 7. The operating conditions of the cathode blanks are aggravated due to the fact that the protective film formed on the metal surface is exposed to mechanical action by the periodically stripped layer of nickel.

Before the operation of nickel deposition, titanium specimens were held in a 260 g/l NaCl solution (pH 2) containing 5×10^{-3} mol $Na_2Cr_2O_7$ at a temperature of 90°C for 10–60 min without a current, or for 1–20 min under cathodic polarization by current densities of 0.1 or 1 mA/cm². NaCl was added to the solution to increase its conductivity. Nickel was deposited at a current density of 50 mA/cm² over 3 h.

It is seen from Table 11.5, that the titanium treatment, even without a current, significantly reduced the titanium hydrogenation. The treatment under cathodic polarization reduced the hydrogenation by more than one order of magnitude.

Cathodic polarization at a current density of 1 mA/cm² over 5–10 min is the optimal regime of titanium pretreatment before nickel deposition. After exposure of the pretreated specimens, which had 4×10^{-6} g/cm² of adsorbed chromium on their surface during 15–20 min without polarization, or under cathodic polarization in the nickel electrorefining catholyte, no presence of chromium was found in the solution, which confirms the stability of the deposited film. However, as seen from Figure 11.21, the content of chromium on the surface of the specimens diminished with an increasing number of nickel deposition-stripping cycles (curve 1), and after the sixth cycle, this content was no more than 10^{-7} g/cm². 70–80% of the chromium had passed from the titanium specimens to the nickel layer, and the rest was lost, probably, during the process of stripping off the deposited nickel layer. Hence, wear of the chromium-containing film takes place during the process of nickel layer stripping. An increasing in the titanium hydrogenation with an increasing number of "deposition – stripping" cycles (curve 2) indicates that this leads to a reduction in the protective properties of the chromate-containing film. The amounts of hydrogenation of the specimens without the protective film and after the fifth "deposition – stripping" cycle are close to one another.

Table 11.5 Hydrogenation of titanium specimens (g-mol H_2/cm²) during 3 h polarization at current density 50 mA/cm² and temperature 85°C as a function of the previous treatment

Treatment time, min	Type of treatment		
	Without current	At cathodic polarization	
		0.1 mA/cm²	1 mA/cm²
0	4.3×10^{-7}	4.3×10^{-7}	4.3×10^{-7}
1	–	–	7.1×10^{-8}
3	–	1.3×10^{-7}	3.6×10^{-8}
5	–	8.3×10^{-8}	1.1×10^{-8}
10	1.2×10^{-7}	7.2×10^{-8}	1.0×10^{-8}
20	–	4.8×10^{-8}	–
30	6.8×10^{-8}	–	–
60	7.1×10^{-8}	–	–

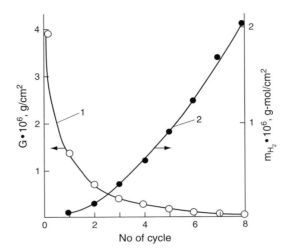

Figure 11.21 Dependences of: 1 chromium adsorption (G) and 2 – hydrogenation of titanium (m_{H2}) on $N°$ of cycle "deposition – stripping" of nickel deposit at cathodic polarization of titanium in catholyte of nickel electrorefining at current density 50 mA/cm². Temperature, 85°C.

Thus, a periodical restoration of the protective film on the titanium surface, by its repeated treatment in the chromate-containing solution, is necessary if this method of protection against hydrogenation is used. In order to diminish the hydrogenation by a factor of 6–7 times, the repeat treatment has to be done after each 2–3 cycles of "deposition – stripping" of the nickel layer. This operation can be included in the technological process of electrochemically refined nickel production. It should be noted that the treatment of titanium in a dichromate-containing solution does not influence the current efficiency of nickel deposition, which is equal to $(95 \pm 1)\%$, as in the case of the deposition on titanium without pretreatment. At the same time, preliminary tests have shown that the introduction of 4×10^{-4} mol $Na_2Cr_2O_7$ into the solution (in accordance with the method of hydrogenation inhibition proposed in Ref. [48]) leads to a considerable reduction in the current efficiency of nickel deposition.

Thus, cathodic polarization of titanium in chromate-containing solutions significantly increases its resistance against hydrogenation and corrosion.

REFERENCES

1. L. M. Yakimenko, Electrode Materials in Applied Electrochemistry (in Russian), Khimiya, Moscow, 1977, 264p.
2. K. Subramanian, V. Arumugam, K. Asokan et al., Central Electrochem. Research Institute, Bulletin of Electrochemistry, Karaikudi, India, Vol. 6, No. 3, 1990, 281–282.
3. L. M. Elina, V. M. Gitneva, V. I. Bistrov and N. M. Shmigul', Elektrokhimiya, Vol. 10, No. 7, 1974, 68–70.
4. V. N. Antonov, V. I. Bistrov, V. V. Avksentiev et al., Khim. Promishlennost', No. 8, 1974, 600–603.

5. A. F. Mazanko, V. L. Kubasov and V. B. Busse-Machukas, Application of metal-oxide anodes in producing of chlorine, caustic soda and oxygen-chlorine compounds. Proceedings of Fourth All-Union Conference "Dimensionally Stable Anodes and their Application in Electrochemical Processes", NIITEKhim, Moscow, 1984, pp. 3–5.

6. I. V. Riskin, M. Kadraliev, G. P. Tutaev et al., Application of Graphite for Producing Current Leak-Offs, Collected articles: "Equipment, its Operation, Repairing and Corrosion Protection in Chemical Industry (in Russian) NIITEKhim, Moscow, Vol. 4, 1976, pp. 6–10.

7. F. I. Mulina, L. I. Krishtalik and A. T. Kolotukhin, Jurn. Prikladnoy Khimii, Vol. 38, No. 12, 1965, 2808–2827.

8. Ya.M. Kolotirkin and D. M. Shub, Today State and Future of Investigations of Anode Materials of Oxides of Base Metals (in Russian), Review "Itogi Nauki i Tekhniki", Electrokhimiya, VINITI, Moscow, Vol. 20, 1982, pp. 3–43.

9. V. L. Mosckalevich, N. P. Juk and E. A. Kalinovsky, Dimensionally Stable Anodes for Aqueous Solutions Electrolysis (in Russion), Collected articles "Khlorine Producing Industry", NIITE-Khim, Moscow, Vol. 12, 1976, pp. 12–13.

10. G. N. Kokhanov, V. V. Avksentiev and R. A. Agapova, Electrokhimiya, Vol. 10, No. 1, 1979, 30–33.

11. I. V. Riskin, G. N. Kokhanov, M. I. Kadraliev et al., Author's Certificate No. 645985, Bulletin of Inventions No. 44, 1980.

12. Yu.V. Baymakov and A. I. Jurin, Electrolysis in Hydrometallurgy (in Russian), Metallurgia, Moscow, 1977, 336p.

13. R. U. Bondar', A. A. Borisova and E. A. Kalinovsky, Elektroknimiya, Vol. 10, No. 1, 1974, 44–48.

14. N. P. Fedotiev, A. F. Alabishev, A. P. Rotinian et al., Applied electrochemistry (in Russian), Khimiya, Leningrad, 1967, 600p.

15. M. A. Dasoyan and I. A. Aguf, The Modern Theory of Accumulator (in Russian), Energiya, Leningrad, 1975, 312p.

16. A. M. Sukhotin ed., Electrochemistry Handbook (in Russian), Khimiya, Leningrad, 1981, 486p.

17. M. Hayes and A. T. Kuhn, Journ. of Applied Electrochemistry, Vol. 8, No. 4, 1978, 327–332.

18. K. Schwabe and R. Radelia, Werkstoffe und Korrosion, Bd. 13, No. 5,1962, ss. 281–284.

19. G. N. Trusov, G. G. Kossy, V. S. Mikheev and B. A. Goncharenko, Author's Certificate No. 505751, Bulletin of Inventions No. 9, 1976.

20. G. G. Kossy, G. N. Trusov, B. A. Goncharenko and V. S. Mikheev, Zashchita Metallov, Vol. 14, No. 6, 1978, 662–666.

21. N. D. Tomashov, V. I. Kazarin, V. S. Mikheev and B. A. Goncharenko, Zashchita Metallov, Vol. 12, No. 5, 1976, 537–540.

22. N. I. Lazarenko, Change in the Initial Properties of the Cathode Surface under the Action of Spark Electric Pulses Passing in a Gas Phase, Collected articles "Metals Treatment in Electric Spark (in Russian) ANSSSR, Moscow, Vol. 1, 1957, pp. 70–94.

23. A. E. Gitlevich, V. V. Mikhaylov, N. Ya. Parkansky and V. M. Revutsky, Alloying of the Metallic Surfaces in Electric Spark (in Russian), Shtiintza, Kiev, 1985, 196p.

24. V. V. Kudinov, Plasma Coatings (in Russian), Nauka, Moscow, 1977, 184p.

25. V. I. Kostikov and Yu.A. Shesterin, Plasma Coatings (in Russian), Metallurgiya, Moscow, 1978, 159p.

26. B. R. Lazarenko, A. E. Gitlevich, S. P. Fursov et al., Electronic Treatment of Metals, No. 1, 1974, 29–32.

27. B. K. Vulf, Thermal Treatment of Titanium Alloys (in Russian), Metallurgiya, Moscow, 1969, 375p.

28. A. E. Gitlevich, V. V. Mikhailov et al., Alloying of Metallic Surfaces in Electric Spark (in Russian), Institute of the Applied Physics of Moldavian Academy of Sciences, Kishinev, Shtiintza, 1985, 196p.

29. M. V. Nevitt, Different Structures with Constant Stoichiometry, in Monograph "Intermetallic Compounds" (in Russian), Metallurgiya, Moscow, 1966, pp. 162–167.

30. N. D. Tomashov, T. N. Ustinova, G. M. Plavnik et al., Zashchita Metallov, Vol. 21, No. 3, 1985, 367–371.

31. I. V. Riskin, A. E. Gitlevich, V. V. Mikhaylov and V. A. Timonin, Electronic Treatment of Metals, No. 6, 1978, 25–28.
32. A. I. Manokhin, B. S. Mitin, V. A. Vasil'eva and A. V. Reviakin, Amorphous Alloys (in Russian), Metallurgiya, Moscow, 1984, 160p.
33. N. D. Tomashov, I. B. Skvortzova, P. B. Budberg et al., Zashchita Metallov, Vol. 19, No. 3, 1983, 398–400.
34. E. W. Justy, H. H. Ewe, A. N. Saridakis et al., Energy Conversion, 1970, Vol. 10, No. 4, 183–187.
35. I. V. Riskin and V. A. Timonin, Zashchita Metallov, Vol. 15, No. 4, 454–456.
36. I. V. Riskin, V. A. Timonin, A. E. Gitlevich and V. V. Mikhaylov, Zashchita Metallov, Vol. 18, No. 3, 1982, 410–413.
37. I. V. Riskin, G. Sh. Gagua and M. L. Bogoyavlenskaya, Zashchita Metallov, Vol. 18, No. 5, 1982, 774–777.
38. N. N. Morar', A. E. Gitlevich, V. V. Mikhaylov and I. V. Riskin, Electronic Treatment of Metals, No. 5, 1983, 23–27.
39. I. V. Riskin, N. A. Zueva and M. L. Bogoyavlenskaya, Zashchita Metallov, Vol. 24, No. 3, 1988, 462–466.
40. I. V. Riskin, V. A. Timonin, A. E. Gitlevich et al., Author's Certificate No. 783365, Bulletin of Inventions No. 44, 1978.
41. I. V. Riskin, V. B. Torshinx, Ya.B. Skuratnik and M. A. Dembrovsky, Materials Performance, Vol. 22, No. 8, 1983, 9–11.
42. I. V. Riskin, V. B. Torshin, Ya.B. Skuratnik and M. A. Dembrovsky, Corrosion, Vol. 40, No. 6, 1984, 266–271.
43. I. V. Riskin and V. B. Torshin, Author's Certificate No. 782416, Bulletin of Inventions No. 14, 1983.
44. V. V. Tzodikov, V. A. Danilkin, L. M. Yakimenko and R. I. Malkina, Zashchita Metallov, Vol. 8, No. 4, 1972, 446–448.
45. L. M. Elina and V. M. Gitneva, Khim. Promishlennost, No. 10, 1974, 775–776.
46. A. M. Sukhotin, Electrochemistry of passivating films on metals. Proceedings of the First Soviet-Japanese Seminar on Corrosion and Protection of Metals, Nauka, Moscow, 1979, pp. 65–85.
47. E. S. Ivanov and N. G. Kliuchnikov, Zashchita Metallov, Vol. 5, No. 5, 1969, pp. 547–550.
48. M. L. Elina, V. I. Eberil', V. M. Yukhim and T. A. Chursina, Author's Certificate No. 283773, Bulletin of Inventions No. 1, 1970.

CHAPTER 12

INDUSTRIAL TESTS AND THE INTRODUCTION INTO ELECTROCHEMICAL PLANTS OF THE DEVELOPED METHODS FOR THE PROTECTION OF METALS AGAINST CORROSION ATTACK BY LEAKAGE CURRENTS

Contents

12.1. PROTECTION BY SACRIFICIAL ANODES IN ELECTROREFINING PLANTS

12.1.1. Industrial tests

Industrial tests of protection from electrocorrosion by dissolving anodes, which drain off the major part of the external anodic current attacking metallic structures, were carried out in a copper electrorefining plant. A corrosion protection system was tested for valves and piping parts made of titanium and stainless steel 18-10.

Anodes for the protection were made of the same copper that was undergoing electrorefining and had the following composition (% weight): 99 Cu; 0.65Ni; 0.1–0.2 O_2; the remainder – Se, Pb, Zn.

In the first stage, tube-shaped anodes (length 15 cm, diameter 18 cm, wall thickness 3.5 cm and weight ~20 kg) were tested. The anodes were mounted at the areas of steel 18-10 pipe to be protected (and in electrical contact with them), between the pipe and a polyethylene pipe. As a result of dissolution by the attack of the anodic current, the anodes were perforated during 20 days of testing. The wall thickness went down from the middle part of the anodes toward their ends.

Weight losses after 20 days of testing were equal to 7.6–8 kg, or, in current units, 13.5–14 A, which corresponded to the measured sizes of the leakage currents. The perforations appeared, first of all, at the ends of the anodes that were adjacent to the polyethylene pipes, where the currents were at a maximum. Contact corrosion on the opposite ends of the anodes resulted from the difference between the open circuit potential of stainless steel 18-10 (0.7 V) and that of copper (0.37 V) in the copper electrorefining electrolyte. The surface area of the steel 18-10 piping was greater than the surface area of the anodes by several orders of magnitude, and that caused quite intensive contact corrosion of the anodes.

To make better use of the material of the dissolving anodes, the shape of the anodes was changed from tube-shaped to rod-shaped (Figure 12.1). At the area of the titanium or steel 18-10 pipe (1) to be protected, four bosses (2), regularly distributed over the pipe perimeter, were welded to the inner pipe surface. The bosses had blind screw holes into which were mounted cantilever copper rods (3) that were 75 cm long and weighed 3 kg. The anodes were oriented along the field of the external current and were located inside the polyethylene pipe (4), which was, as in the previous case, a continuation of the metallic pipe. The thread which connected the anodes with the metallic pipe, and which provided electrical contact between them, was sealed with gaskets (5).

Tests lasting 85 h showed that the total weight losses of four rod-shaped anodes were 1–3 kg, which corresponded to current sizes of 10–30 A. As a result of the attack by an external anodic current, the ends of the anodes were sharpened (Figure 12.2a). The weight losses of the anodes that were under the action of external cathodic currents were relatively small (Figure 12.2b).

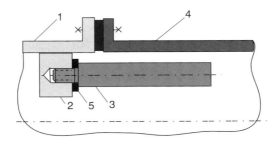

Figure 12.1 Mounting of rod-shaped copper anode inside a protected pipe (explanations are given in the text) (see color plate 11).

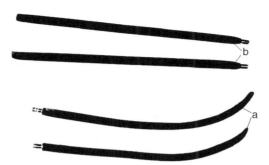

Figure 12.2 Dissolving rod-shaped copper anodes after 3.5 days of testing in technological piping of a copper electrorefining plant at the areas of action of current leakages: a – anodic; b – cathodic.

The results of the tests indicate that the dissolution rate of the anodes is determined by the external anodic current. Using rod-shaped anodes mounted in the cantilever enables more efficient use of the anode material, which dissolves preferentially at the free end of the rod, where the current is at a maximum.

However, using rod-shaped anodes in contact with the protected metal inside the electrolyte does not solve the problem of contact corrosion. It is seen in Figure 12.2b that in the areas adjacent to the screw, the rods undergo noticeable corrosion. This may lead to a failure of the electric contact in the screw connection. Thus, it is necessary to avoid locating the electric contact inside the electrolyte, which is often difficult to do.

Moreover, the possibilities of internal surface protection of a pipe with the help of dissolving anodes are limited by the following factors: increase in the hydraulic resistance of the solution flow inside the pipe and the necessity of controlling and frequent replacement of the anodes. Protection by dissolving anodes is most feasible when it is used for objects on which the contact with the anodes can be kept out of the aggressive medium, as controlling and replacing the anodes can be easily carried out during the operation of the plant, without shutting it down.

12.1.2. Introduction of the protection in a copper electrorefining plant

Degassing tanks, which are used in copper electrorefining plants to reduce the saturation of the electrolyte with gases, can be effectively protected from electrocorrosion with the help of dissolving anodes [1, 2]. The protected tank 1 (Figure 12.3a) had the shape of a parallelepiped, and was filled with the electrolyte to more than half its height. The anode 2 was hung up by ears, which were introduced into slits made in the upper part of the tank walls (Figure 12.3a and b), above the level of the electrolyte. The dimensions of the tank allowed standard 240 kg anodes to be placed inside it and subjected to dissolution in the electrolytic bath (with subsequent deposition of the refined copper on the cathode blanks). Removal of the remains of the dissolved anodes and installation of new anodes into the tanks was carried out during the periods of cleaning the baths, about once every 20 days.

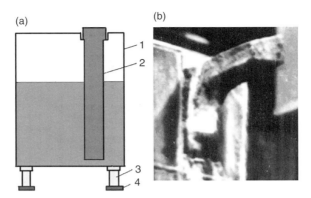

Figure 12.3 Titanium degassing tank (1) with protection from corrosion by anodic leakage current with the help of a dissolving anode (2): **a** – design of the tank with anode inside (explanations are given in the text); **b** – photograph of the upper part of the tank with a slit in which an ear of the anode is introduced (see color plate 12).

It was noted in Chapter 6 that under the attack of a leakage current, the bottoms of the degassing tanks are primarily damaged in the welding seam zones, which coincide with the areas of leakage current attack. It was shown that welding drastically reduces the corrosion resistance of titanium in the copper electrorefining electrolyte under anodic polarization. In this connection, steps were taken to remove welding seams from the areas of attack by leakage currents. For this purpose the tanks were installed on welded 20 cm-high titanium supports (3) (Figure 12.3a) provided with insulators (4). No cases of corrosion damage to the tanks provided with this means of protection were recorded over 3 years of their operation.

12.2. INDUSTRIAL TESTS AND THE INTRODUCTION OF CORROSION PROTECTION TO TITANIUM STRUCTURES BY CURRENT LEAK-OFF ANODES

12.2.1. Industrial tests

In Chapter 10 it was shown that titanium, which is characterized by the relation $\Delta = (E_{ox} - E_a) < 0$, can be protected from corrosion attack from external anodic currents by stable current leak-off anodes. Testing this protection principle under industrial conditions was carried out on titanium group headers for wet chlorine, and on titanium electrolyzer covers in chlor-alkali diaphragm electrolysis plants. On the headers, the anodes were installed on the branch tubes, at the side of the negative rectifier pole, and on the covers, on the nozzles at the side of the positive pole. The anodes were preferentially installed at the branch tubes disposed above the marginal electrolyzers in a set. Rubber-lined and faolite tubes for chlorine withdrawal (chlorine taps) connected the electrolyzers with the group headers. As already noted, leakage currents along these tubes are the highest, due to

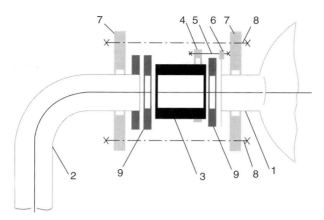

Figure 12.4 Protection of titanium branch tubes of group headers for wet chlorine with the help of current leak-off anodes of graphite (explanations are given in the text) (see color plate 13).

chlorination of the rubber and faolite. Thus, the testing conditions were very severe from the point of view of attack by anodic leakage currents. Anodes of graphite MG, non-thickened and thickened by pyrolytic carbon, and dimensionally stable anodes with coatings of ruthenium–titanium oxide (ORTA) and of manganese–cobalt oxides (OMCA) were tested.

Tube-shaped graphite anodes (3) (Figure 12.4) were installed between the branch tubes (1) to be protected and the chlorine taps (2) made from insulating material. The anodes were 50 mm long and had an internal diameter of 50 mm (corresponding to the diameter of the branch tube) and a wall thickness of 25 mm. Electrical contact between the anodes (3) and the branch tubes (1) was provided through metallic clamps (4) placed on the anode and bolts (5) that connected the clamps (4) with the flanges (6) of the titanium branch tubes (1). The assemblies were fixed by mounting flanges (7) and tie bolts (8), and sealed by gaskets (9).

During the first 3 months of testing, the internal surface of all three tested anodes of non-thickened graphite loosened to a depth of about 10 mm, and the anodes were completely destroyed after the subsequent 5 months of operation. The anodes of thickened graphite corroded at their lower part (where the condensate accumulated) to a depth of about 10 mm after 8 months of testing. Thus, their resistance increased; however, a perforation occurred on one of the tested anodes during this time. The titanium branch tubes protected by these anodes did not undergo corrosion damage.

It is impossible to prevent destruction of the graphite caused by the preferential oxygen evolution in dilute chloride solutions under the attack of anodic leakage currents. Therefore, dimensionally stable anodes of titanium with metal-oxide coatings were tested in the next stage. The design of the tested anodes was similar to the design of titanium "sacrificial inserts" shown in Figure 9.8. The ORTA and OMCA coatings were applied to the internal surface of the anodes and to the surface of the flanges which were in contact with the gaskets. Due to these electrochemically active coatings, the inserts functioned as current leak-off anodes.

No damage occurred at the 10 anodes of ORTA after 8 months testing. However, one of these anodes was perforated after a year and a half and another anode, after 2 years of operation. Both of them were located above the marginal electrolyzers in a series (where the magnitudes of the leakage currents are the highest). Damage to the coating and pitting of the titanium base was noted at some areas of other anodes.

Damage to the OMCA anodes located above the marginal electrolyzers in a series was observed after 4–6 months of testing.

In all the considered tests, the protected branch tubes did not undergo corrosion damage.

Thus, the considered principle provides effective protection against corrosion by an external anodic current in the wet chlorine lines. The duration of the protection depends on the lifetime of the current leak-off anode. This lifetime can be increased by reducing the magnitudes of the leakage currents.

On the titanium covers (1) of diaphragm electrolyzers (Figure 12.5), alongside the nozzles for connecting the chlorine withdrawal tubes, areas of brine input into the electrolyzers were damaged by anodic leakage currents. These areas include conic nozzles (2) with rubber plugs (3). The ends (5) of the brine hoses (4) pass through these plugs. These assemblies belong, in principle, to the brine lines. Current leakages along the brine hoses, as shown in Chapter 3, surpassed 1 A.

The conic nozzles (2) are usually welded to the covers; however, in order to apply a coating on their surface, they are made to be removable. A ruthenium–titanium oxide coating (ORTA) which is indicated in Figure 12.5 by red lines, was applied to the internal surface and the flanges of the current leak-off anodes and the conic nozzles. Thus, the conic nozzles (2) were protected from electrocorrosion by the electrochemically active coating and at the same time, played the role of current leak-off anodes for the adjacent areas of the titanium covers. Only ORTA coating was tested on the conic nozzles, because of the high values of the current densities in the brine lines.

The tests proved that the protection of titanium covers from electrocorrosion by the considered current leak-off anodes was effective. Neither the nozzles nor the adjacent areas of the covers were damaged during 1 year of testing.

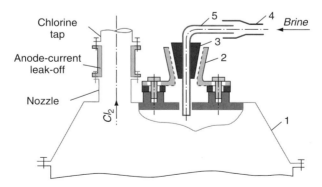

Figure 12.5 Protection of nozzles for connection chlorine taps to headers and for brine input mounted on titanium covers of diaphragm electrolyzers (explanations are given in the text) (see color plate 14).

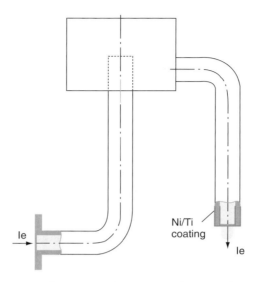

Figure 12.6 Titanium overflow unit for caustic liquor of diaphragm electrolyzer provided with protection from electrocorrosion by Ni/Ti coating obtained in an electric spark explanations are given in the text (see color plate 15).

The results obtained in these tests in the areas of brine input into the electrolyzers showed that effective protection from corrosion by anodic leakage currents in the brine lines of other types of titanium equipment can also be executed by ORTA anodes. These results were also extended to the protection of titanium equipment in the chloranolyte lines in mercury cathode electrolysis plants.

Protected titanium overflow units were tested in the caustic liquor lines in diaphragm electrolysis. Titanium with a Ni/Ti coating obtained in an electric spark was used as material for the current leak-off anodes. The end areas of 50 mm diameter tubes were attacked in these units by anodic leakage currents (Figure 12.6). A Ni/Ti coating was applied, with the help of a mechanized installation for coating in an electric spark, to the internal and external surfaces of these areas for a length of 80 mm. The coated areas played the role of the current leak-off anodes. Any damage to the protected units was not fixed over 1 year of operation. The end areas of such units were usually destroyed after less than half a year in the absence of the protection means.

It was noted in Chapter 5 that carbon steel corroded under anodic polarization in caustic liquor. A potential of 0.73 V was established on carbon steel at a current density of $20 \, mA/cm^2$, and its corrosion rate was equal to $1.6 \, g/m^2 h$. During the polarization of a combined electrode of carbon steel and titanium with a Ni/Ti coating at a current density of $20 \, mA/cm^2$ (based on the total surface of the combined electrode) the corrosion rate was equal to $1.5 \, g/m^2 h$, i.e., it practically did not go down. This is explained by the fact that the potential values established on the carbon steel and titanium with the Ni/Ti coating were close, under the considered conditions; therefore, a significant share of the external current was concentrated on the carbon steel portion of the combined specimen.

Apprehensions arose in this connection raised concern that the frames of the electrolyzers contacting with the overflow units may be damaged by anodic leakage currents. Therefore, a calculation was made of the potential drop along the tubes of the overflow unit from the end protected by the Ni/Ti coating towards the cross-section of its contact with the frame. The following parameters were taken into account in this calculation, which was done with the help of equations (8.6) and (8.7) for a monopolar tube: length of tube area with Ni/Ti coating – 8 cm; total length of titanium tubes without coating – 135 cm; radius of tubes 2.5 cm; solution conductivity – 1.1 ohm cm. The slopes of the polarization plots used in the calculation for the Ni/Ti coating and for the titanium were equal, respectively, to 0.075 and 0.4 V (see Figures 6.2 and 11.14). The measured size of the leakage current used in the calculation was equal to 0.4 A.

The calculation showed that in the cross-section of the contact of the overflow unit with the frame of the electrolyzer, the potential goes down to 0.11 V. 5 h long potentiostatic tests did not reveal any corrosion of carbon steel specimens at this potential value.

Industrial tests confirmed the results of the calculation and of the laboratory test. Measurements carried out during the work process have shown that in the cross-section of the frame contact with the overflow unit, the potential was equal to 0.07 V, i.e., it was quite close to the calculated potential value. Tubular inserts 10 cm in length and with a radius of 2.5 cm installed between the overflow units and the frames of the electrolyzers did not corrode during 1 year of testing. Thus, corrosion of the frames was prevented due to the reduction of the potential along the tubes of the overflow units. The anode with the Ni/Ti coating is oriented along the field of the external current with respect to the frame of the electrolyzer. The titanium tubes of the overflow unit place this anode at a remote distance with respect to the frame, along the field of the leakage current.

Destruction of the Ni/Ti coating was noted at the end areas of the overflow units that were installed on the side of the negative pole of the power source, where these areas were attacked by cathodic leakage currents. However, the units themselves did not corrode. The possible causes of the destruction of the coating by cathodic currents were considered above.

The data on the destruction to the Ni/Ti coating show that the parts of metallic structures that are protected by this coating against corrosion by external anodic currents in a series of electrolyzers, cannot be transferred to the zones where the action of a cathodic current is possible. In particular, when the protected overflow units are installed, they must be marked in accordance with the number of the particular electrolyzer. After any temporary removal, they must be returned to the same electrolyzers.

12.2.2. Introduction of protection

Protection of titanium piping and equipment from corrosion by external anodic leakage currents with the help of current leak-off anodes has found extensive application, initially, in chlor-alkali electrolysis plants [3], owing to its simple technical implementation and high efficiency.

Headers for wet chlorine in diaphragm and mercury cathode chlor–alkali electrolysis plants were the first targets of this method of protection. The anodes were installed at the areas of current interruption along the titanium group and set headers by inserts of insulating material, and at the branch tubes for mounting tubes of insulating materials that connected the titanium group headers for wet chlorine with the electrolyzers.

As corrosion damage of some of the current leak-off anodes occurred in the industrial tests, means for reducing the magnitude of the leakage currents from the electrolyzers were taken, alongside the protection by current leak-off anodes. Chlorine taps of Teflon for connecting the electrolyzers with the wet chlorine headers were installed instead of chlorine taps of rubber–lined carbon steel or faolite. It was noted that such a substitution drastically reduced the leakage currents, but it may lead to a change in the leakage current direction which cannot be controlled. Therefore, the anodes were installed above all the electro-lyzers, irrespective of their position in the set of electrolyzers. Thus, they were installed in the branch tubes of the group headers on both sides of the zero point in a set of electrolyzers. Photographs of a diaphragm electrolysis plant are given in Figures 12.7a and b, a general view and the area of the titanium group header for wet chlorine with current leak-off anodes installed at the tube branches. Teflon chlorine taps are connected to the other side of these anodes. Both ORTA and OMCA anodes were applied for protection.

More than 10 years experience in the protection from electrocorrosion of titanium headers for wet chlorine by current leak-off anodes, in four chlor–alkali electrolysis plants, has proved its efficiency. None of the protected headers has undergone corrosion damage.

The titanium covers for diaphragm electrolyzers were protected at two areas of leakage current attack, as indicated in Figure 12.5. Protection of the titanium nozzles connecting the electrolyzer covers with the chlorine taps was similar to the protection of the branch tubes of the group headers. However, unlike the headers, the nozzles of the covers that are attacked by anodic leakage currents are disposed at the side of the positive, and not the negative, pole of the power source in a set of electrolyzers (see Figure 3.2).

Using Teflon chlorine tap, provided a reduction in the current at both cross-sections of their connection: at the branch tubes of the titanium headers and at the nozzles of the titanium covers. On both of these areas identical current leak-off anodes were installed that had either ORTA or OMCA coatings [4]. As noted above, only ORTA anodes were used for coating the conic nozzles for the brine input into the electrolyzers.

5–10 years long operating experience of the protection of titanium covers by the anodes long; this confirmed the efficiency of this type of protection.

Owing to the rise in the cost of titanium, producers of equipment for electro-chemical plants have developed constructions for the covers for electrolyzers which allow a saving in the titanium consumption. Thus, in patent [5] molding the covers of polymeric materials, and lining their inner surface with a thin, in the order of tenths of a millimeter, layer of titanium is proposed. The lining is primarily formed in accordance with the shape of the polymeric part of the cover. Since the

(a)

(b)

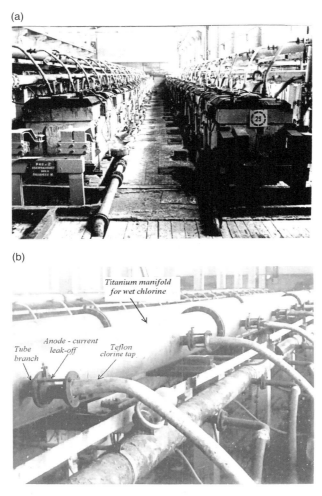

Figure 12.7 a – General view of diaphragm electrolysis plant with titanium header for wet chlorine protected by current leak-off anodes; b – an area of the protected header.

titanium construction practically does not corrode when it is protected by a current leak-off anode, in combination with a means of current reduction, the thickness of the lining layer does not influence its protective properties, and such covers may have a long lifetime. It should be noted that lining polymer piping with titanium protected by the considered system, should drastically reduce the titanium consumption on the wet chlorine lines.

Corrosion of titanium shell-and-tube heat exchangers (brine heaters) by anodic leakage currents occurred in two diaphragm electrolysis plants (1) (Figure 12.8). Tubes (2) of the heat exchanger corroded at the areas of their expansion inside the tube plate (3). Therefore current leak-off anodes (4a) and (4b) were installed just in

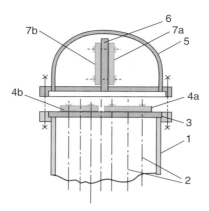

Figure 12.8 Protection of titanium shell–and–tube heat exchangers (brine heaters) by current leak–off anodes (explanations are given in the text) (see color plate 16).

these areas. They were cut from titanium sheets and had the form of semicircles with the same radius as the radius of the tube plate of the heat exchanger. Holes with a radius equal to the average radius of the tubes were made in the anodes. The holes were distributed in accordance with the distribution of the tubes over the tube plate (3) so as not to hinder the flow of the solution inside the heat exchanger.

Two anodes (4a) and (4b) were mounted on a tube plate and covered almost the entire surface of the tube plate, as shown in Figure 12.8. On radial partitions (6) of the covers (5) of the heat exchanger, i.e., also in the action zone of leakage currents, were mounted two additional anodes (7a) and (7b), in the form of rectangular plates, to reduce the density of the leakage currents. Each anode covered about 80% of the partition surface. All the anodes were fixed with the aid of titanium bolts, which provided electrical contact of the anodes to the heat exchanger. Leakage currents acting on the brine heaters are usually low, because these apparatuses are located at a distance from the electrolyzers. Therefore, OMCA current leak-off anodes were used for their protection. The success of this protection was confirmed by 7 years operation without any signs of corrosion damage.

The relatively small dimensions of valves and other kinds of accessories enable the introduction of these objects into stoves with regulated temperatures, for applying the electrochemically active coatings by thermal decomposition methods. Thus, metal-oxide coatings can be applied directly to the areas of these accessories that are attacked by external anodic currents. These areas of limited dimensions play the role of current leak-off anodes.

Protection of 100, 150 and 200 mm diameter valves was carried out on the brine lines of two diaphragm electrolysis plants. As the leakage currents along these lines are high, more than 10 A, an ORTA coating was used for the protection. It was applied to the flanges, and areas adjacent to the flanges of the valves, as well as to the non–working surface areas of the rods (Figure 12.9). Valves with this protection operated without any corrosion damage for more than 7 years.

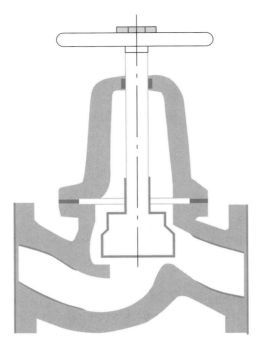

Figure 12.9 Titanium valve provided with protection against electrocorrosion by ORTA coating indicated by red lines (see color plate 17).

Units for drainage of the current from the electrolyte to the earth must be installed on the electrolyte lines, particularly, on the brine lines in electrolysis plants [6, 7], when the piping of these lines is made of insulating material. Units of graphite are not suitable because of the low mechanical strength of graphite. An attempt to apply tube-shaped flanged bobbins of carbon steel on a brine line for this purpose, also failed. Bobbins which had a wall thickness of 15 mm were completely destroyed within 2–3 days of operation, as a result of attack by anodic leakage currents, which surpassed 10 A on this line.

Units of carbon steel were substituted by units of titanium with an ORTA coating. They were installed at each group header, close to the set header (Figure 12.10a). Both set and group headers for the brine were made of rubber–lined carbon steel. The coating was applied to the internal surface and to the flanges of each bobbin (Figure 12.10b). As the current density on the surface of these units was several orders of magnitude lower than the current density on the anode surface in the industrial electrolysis, only three layers of ORTA were applied to the titanium surfaces of the units (instead of the four layers that must be applied to the surface of the commercial anodes in accordance with the industrial standards). The units were successfully used in two diaphragm electrolysis plants. At none of the group headers did corrosion damage of the units occur during more than 8 years of their operation.

(a)　　　　　　　　　　　　(b)

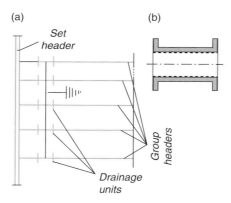

Figure 12.10 a – Scheme of installation of units for current drainage from brine to earth for group brine headers; b – unit of titanium with ORTA coating indicated by dotted lines (see color plate 18).

12.3. INDUSTRIAL TESTS AND THE INTRODUCTION OF CORROSION PROTECTION BY ANODES ORIENTED ALONG THE FIELD OF THE EXTERNAL CURRENT IN ELECTROCHEMICAL PLANTS

12.3.1. Industrial tests

Industrial tests of protection by anodes oriented along the field of the external current were carried out in a copper electrorefining plant at two areas of stainless steel 18–10 piping of 15 cm diameter. It was shown above, that in the technological solution of this plant, as opposed to titanium, stainless steel 18–10 is characterized by the relation $\Delta = (E_{ox} - E_a) > 0$ when lead anodes are used for protection against corrosion by anodic leakage currents.

A tube-shaped lead anode (1) (Figure 12.11) of 150 mm diameter and 600 mm length, was installed at the area to be protected, inside the tube (2) of insulating material (polyethylene) adjacent to the metallic piping (3). Remote placement of the anode along the direction of the external current, and contact with the metallic pipe, were provided by a 2 m long interstitial titanium tube (4) of diameter 150 mm, and by a 0.5 m long titanium valve (5) (used for technological

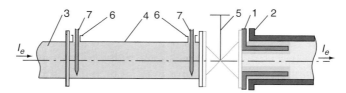

Figure 12.11 Scheme of protection of stainless steel 18–10 piping in technological solution of copper electrorefining with the help of a lead anode oriented in the field of the external current (explanations are given in the text) (see color plate 19).

Table 12.1 Potential values at the terminal areas of titanium tubes in the protection systems of two steel 18-10 pipes with the help of anodes oriented in the field of the external current

No. of series	Average value of leakage current (A)	Potentials at the cross-section of titanium tube siding:	
		to the valve	to the SS 18–10 piping
1	16.7	1.00–1.05	0.70–0.75
2	20.0	1.10–1.15	0.80–0.90

reasons), which were installed between the metallic pipe and the anode. Thus, the total distance of the remote placement of the anode, including the length of the anode itself, was 3.1 m. Nozzles (6) were made at the terminal areas of pipe (4) for installing Teflon capillaries (7). The capillaries were connected to reference electrodes through electrolytic bridges (not shown in the figure). Potentials were measured in these areas by a millivoltmeter of high input resistance.

As can be seen from the results of the measurements given in Table 12.1, the potential along the titanium tubes goes down from the value of the transpassivation potential to the value of a stable passive state. Slightly more positive potential values in series 2 correlate with higher magnitude leakage currents.

After 1 year of testing, no signs of corrosion damage were revealed on either of these pipes, whereas under similar conditions, unprotected 18-10 stainless steel pipes were often perforated at the areas of leakage current attack during this period.

Thus, corrosion tests, in combination with measurements of potentials and currents, confirmed the efficiency of the protection of metals that are characterized by the relation $\Delta > 0$, with the help of stable anodes oriented and placed remotely along the field of the external current.

12.3.2. Introduction of protection

Based on the results of the considered industrial tests, protection of stainless steel 18-10 piping against attack by external cathodic currents was introduced into the copper electrolysis plant. A 300 m length pipe line that consisted of 15 cm diameter pipes, was protected [8]. The dimensions of the lead anodes and the interstitial titanium tubes installed at the protected areas in the industrial tests and in the final introduced protection system were similar.

Measures were taken to remove the welding seams from the zones of leakage current action, because stainless-steel welds are prone to intercrystalline corrosion in acid–sulfate solutions, even in the absence of an external current. The construction of the flange connections excluded any contact of the welding seams with the electrolyte.

The stainless steel piping provided with this protection system operated for 5 years without any signs of corrosion.

12.4. Industrial Tests of Titanium Protection Against Corrosion Attack by Cathodic Leakage Currents With the Help of Stable Cathodes

Industrial tests of stable titanium cathodes were carried out in the chloranolyte line in a mercury cathode chlor-alkali electrolysis plant. The area of a pipe in the titanium header adjacent to an insulating insert served as a stable cathode.

The accumulation of oxygen–chlorine compounds near the protected area was provided by installing inside the said pipe area a perforated Teflon sleeve that formed an annular gap – stagnant zone between the walls of the protected pipe and the sleeve. In the lower part of the protected pipe was installed a nozzle for sampling the solution from the annular gap.

No corrosion damage on the protected pipe was revealed during 1 year of testing. Before the installation of the considered protection means at this area, which was attacked by cathodic leakage current, pipe perforation occurred during six months of its operation.

Analysis of the solution samples that were taken from the annular gap showed that the minimum content of hypochlorite and chlorate ions in these samples surpassed their content in the bulk of the solution by more than one order of magnitude (Table 12.2). The higher chlorate concentration in the annular gap and in the solution bulk, as compared with hypochlorite, is explained by its higher stability.

It was shown in Chapter 7 that for reducing the corrosion rate of titanium by more than one order of magnitude, i.e., to a non-dangerous corrosion level, it is enough, under the considered conditions, to attain a concentration of 0.3 g/l of one of the anions of the above-mentioned oxygen–chlorine compounds. As seen from Table 12.2, the total concentration of these anions inside the annular gap is much higher.

A growth in the potential on the titanium took place along the protected pipe area disposed in the annular gap. The potential rose in accordance with the direction of the external cathodic current and, as a result of its growth to a non-dangerous value, outside the boundaries of the annular gap where the concentration of the oxygen–chlorine compounds was low, the titanium pipe as a whole did not corrode. Thus, in accordance with the developed concept, the pipe area that was disposed inside the annular gap represented a stable cathode oriented along the field of the external cathodic current.

Table 12.2 Concentration (g/l) of ions ClO^- and ClO_3^- inside the annular gap and in the bulk of the solution

In annular gap		In solution bulk	
ClO^-	ClO_3^-	ClO^-	ClO_3^-
0.25–0.50	0.60–4.00	0.01–0.02	0.06–0.07

12.5. COMBINED PROTECTION OF METALS AGAINST CORROSION ATTACK BY LEAKAGE CURRENTS

When corrosion attack by a cathodic current takes place at some areas of a metallic structure, attack by anodic currents is also usually unavoidable at some other areas of this structure. In these cases, combined protection from corrosion attack by external currents of both the anodic and cathodic directions must be provided.

The action of currents of both anodic and cathodic directions is most typical for the areas of metallic piping at which insulating inserts are installed. As the piping is disposed in the field of the external current, on one side of the inserts the metal is attacked by an anodic, and on another side, by a cathodic current.

Such a combined protection was introduced on titanium headers for the chloranolyte line in a mercury cathode chlor-alkali electrolysis plant. ORTA anodes were installed at the areas of attack by the anodic currents, and perforated Teflon sleeves, which formed stagnant zones, were installed at the areas of attack by the cathodic current (Figure 12.12). This protection was installed at the areas of insulating inserts and at the branch tubes for connecting the group manifolds with the electrolyzers. No cases of corrosion damage of the headers were noted during more than 5 years of operation.

The results of the industrial tests and cumulative experience of the commercial application of the developed methods and means of protection from electrocorrosion show that in most cases these methods provide long-time corrosion stability of the protected piping and equipment of passive structural metals under the conditions of attack by external currents. However, when only electrochemical methods are applied, the protection is not always effective enough. A combination of the different methods and means is often necessary to provide reliable operation of the equipment. In particular, combined protection must be applied when elevated leakage currents destroy the current leak-offs anodes. Protection of titanium group headers for wet chlorine by current leak-off anodes, in combination with a means of reducing the leakage current (by using Teflon tubes for the withdrawal of chlorine from the electrolyzers) can be considered as an example of such a combined protection.

Application of oriented lead anodes in combination with improved distribution rakes which increased the electrical resistance of the electrolyte in a copper

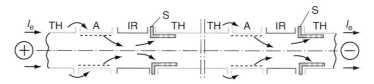

Figure 12.12 Combined protection of titanium chloranolyte header from corrosion by external currents of anodic and cathodic directions: TH – protected header; A – current leak-off anode with ORTA coating (indicated by dotted lines); IR – insulating inserts; S – perforated sleeves of Teflon. Arrows indicate the current direction (see color plate 20).

electrorefining plant, is another example of combined protection. These rakes reduced the leakage currents along the technological lines by half and consequently, increased the lifetime of the lead anodes that protected the steel 18-10 piping and of the dissolving copper anodes that protected the titanium degassing tanks. Construction improvements that enable the removal of the welding seams of the flange connections of the piping and the degassing tanks from the attack zones of leakage currents can also be regarded as additional means of combined protection from electrocorrosion.

Sometimes a simple means such as regular cleaning of the devices for breaking the jet of the alkali liquor at the output of the overflow units and increasing the lengths of the rubber hoses for feeding the brine to the electrolyzers significantly increase the lifetime of the current leak-off anodes.

Rational equipment distribution along the series of electrolyzers, in combination with other necessary protection means, taking into account the electrochemical characteristics of the structural metal of the equipment and the direction of the external current, can also be considered as a combined protection system. Thus, distribution of the titanium and stainless steel 18-10 valves was executed in a copper electrorefining plant. Titanium valves protected by lead current leak-off anodes were installed at the areas of the action of anodic leakage currents. Valves of steel 18-10, without any protection, were installed at the areas of cathodic current action, since steel 18-10 is corrosion-resistant under these conditions. None of the valves was corroded during 5 years of operation. In the former, random distribution of these valves, some of them were destroyed within several months.

Thus, combined protection that takes into account specific features of the working conditions of the equipment and that applies electrochemical methods as a major part of the protection system enables the achievement of a complete solution of the problems of protection against electrocorrosion under the conditions of attack by external anodic and cathodic currents.

12.6.   EFFECTIVENESS AND POSSIBILITIES OF USING THE DEVELOPED METHODS OF METAL PROTECTION AGAINST ELECTROCORROSION

The lifetime of the protected metallic structure, and the expenses incurred for controlling and replacing the protective anodes and cathodes, can be used as the criteria for estimating the effectiveness of the developed electrochemical methods of electrocorrosion protection of these structures in the field of external currents. The passive metal structures are the objects of the protection and the developed methods of protection are based on maintaining the metal in the field of its passive state. Therefore, the corrosion of a well protected structure is insignificant and the structure can remain in a state of operation as long as the protection is provided. Thus, in the case under consideration, the effectiveness of the protection is determined by the lifetime of the anodes and cathodes that are applied in the protection systems.

Anodes that are used in the regarded protection systems can be divided into three major groups: dissolving anodes, anodes of limited corrosion stability and stable anodes. With regard to cathodes, only systems with stable cathodes were developed for protection from corrosion attack by cathodic currents.

The effectiveness of electrocorrosion protection with the help of dissolving anodes depends on the specific features of the equipment operation. Their advantage is the absence of material expenses for producing the anodes, and their limitation involves the necessity of constantly controlling and replacing the consumed anodes. The example given of protecting degassing tanks by dissolving anodes in copper electrorefining plants shows that, when dissolving anodes of a large mass are used, their replacement has to be performed relatively infrequently. The replacement operation can be included in the maintenance schedule and executed at constant intervals, which reduces the maintenance expenses.

The application of anodes made of materials having a limited stability, such as lead and graphite, which are significantly corroded under the attack by anodic currents, may be suitable in cases where their lifetime is long enough and they can be easily replaced after destruction. An estimation of the expediency of their application can be done at the design stage.

It is clear that the methods developed for the electrocorrosion protection of metals with the help of stable anodes and cathodes are the most effective. The advantages that determine their widespread application in the chemical and nonferrous metallurgical industries are as follows:

1. Low labor and expenditure for protection of piping and equipment, since a limited number of areas have to be protected. Thus, for the protection of two titanium headers for wet chlorine in an average diaphragm electrolysis plant consisting of 240 electrolyzers, from 150 to 240 current leak-off anodes are necessary, each of which is 50–100 mm long and has diameter of about 50 mm. The quantity of ruthenium for coating these anodes is $\sim$50 g, i.e., less than the quantity of ruthenium required for a set of commercial anodes used in one electrolyzer.
2. Low maintenance expenditure. Electrochemical protection with the aid of stable anodes is carried out without using external power sources or sacrificial anodes, i.e., without means that require the expenditure of energy and labor for constant control and replacement of the destroyed anodes. Periodic random inspection of the state of the anodes and of their contact with the protected areas is enough to ensure their reliable operation. The same amount of maintenance is sufficient for stable cathodes.

Owing to the low cost, effectiveness and simplicity of implementation of the developed methods of protection from electrocorrosion, and owing to the fact that the unprotected metallic piping and equipment, as a rule, undergo intensive corrosion under attack by external currents, extensive introduction of the considered protection methods is expected in existing and newly designed plants, especially in the chemical and metallurgical branches of industry.

It is of interest to consider the possibilities of using the above-mentioned principles of electrochemical protection for underground and underwater piping and other structures under attack by stray currents. In particular, anodes oriented in the field of an external current may broaden the possibilities of protection by polarized drainage. Under some conditions, electrodes that include combinations of stable anodes and cathodes can be used.

It is worthwhile remembering that the developed principles are oriented to the protection of passive metals. Consequently, a most effective protection of aluminum structures in soil and water, and of passive carbon steel, may be achieved in accordance with these principles. Protection of the carbon steel reinforcement, which under normal conditions has a passive state inside the concrete, against corrosion by stray currents may be of particular interest.

Of course, expansion of the range of applications of the above-considered principles of metal protection from electrocorrosion requires further extensive and comprehensive investigation.

REFERENCES

1. I. V. Riskin, L. M. Lukatsky, I. A. Travnichek et al., Tzvetnaya Metallurgiya, No. 2, 1983, 31–34.
2. I. F. Khudiakov, A. I. Tikhonov, V. I. Deev and S. S. Naboychenko, Metallurgy of Copper, Nickel and Cobalt (in Russian), Part 1, Metallurgiua, Moscow, 1977, 295p.
3. I. V. Riskin, M. I. Kadraliev and G. P. Tutaev, Khim. Promishlennost', No. 7, 1978, 51–52.
4. V. Riskin, M. I. Kadraliev and G. P. Tutaev, Khim. i Neftyanoye Mashinostroenie, No. 12, 1977, 17.
5. Patent No. SE19930003938 19931126, A. Ullman, M. Kroon, S. O. Boquist, M. Karlsson, Permascand AB, Published 15.09.1998.
6. L. M. Yakimenko, Electrolyzers with Hard Cathode (in Russian), Khimiya, Moscow, 1966, 300p.
7. I. A. Kornfeld and V. A. Pritula, Protection of Reinforced Concrete from Corrosion by Stray Currents (in Russian), Stroyizdat, Moscow, 1964, 75p.
8. I. V. Riskin, L. M. Lukatsky and V. B. Torshin, Tzvetnaya Metallurgiya, No. 22, 40–43.

INDEX

Activators:
  bromide, 81
  chlor-ion, 8, 13–15, 25, 71–72, 79,
    81–84, 89–90, 92–94, 96
  iodide, 81
Active state of metals, 9–11, 18
  aluminum, 11–12, 18
  carbon steel, 23, 28, 33, 56–57, 143
  titanium, 80, 148
Adsorption, 1, 215–216
Aluminum, 8, *see also* Active state of metals,
    aluminum
Amorphous structure of titanium, 205–206
Anatase, 80, *see also* titanium oxides
Anodes:
  alternatively dissolving, 159–160, *see also*
    Anodes, dissolving
  ceramic, 34, 60, 179, 191
  current leak off, 169, 176, 224, 236*f*,
    236–238
  dimensionally stable, 165–168
  dissolving, 159–160, 169, 221–223,
    223*f*, 238
  graphite, 147–148, 165–168, 167*f*,
    224–225, 232, 238
  intermetallic, 97–99, *see also* Intermetallic
    coatings on the surface of titanium
    and Ti-0.2%Pd
  lead, 196
  magnesium, 17, *see also* Anodes, sacrificial
  magnetite, 199, *see also* Magnetite
  manganese-cobalt oxide coating,
    193–194, 194*f*, 195*f*, 224–225
  nickel, 168, 176, 178, 199, *see also* Nickel
  OMCA, *see* manganese-cobalt oxide
    coating
  oriented in the field of external currents,
    170–171, 172*f*, 176–178, 184
    extended, 171–172, 171*f*

remotely placed, 170–174, 171*f*,
  176–178, 184, 228, 233–234
ORTA, *see* ruthenium-titanium oxide
  coating
ruthenium-titanium oxide coating, 190,
  192–193, 195
platinum, 164–165, 197–198, 197*f*,
  212–213, *see also* Platinum
sacrificial, 155, 157, 169, 172–173, 225,
  238, *see also* Anodes, dissolving
silicon cast iron, 34
steel scrap, 33–34
zinc, 18, 31–32
Anodic protection, 59, 80, 154
Austenite, 17
Auxiliary electrode, 10, 104, 154

Bacteria activity, 24
Batteries, 1, 39
Barrier film forming, 15, 193
Barrier film on titanium, 81–83, 97, 99,
  164, 167, 193, *see also* Titanium, oxides
Bus ducts, 32

Carbonic acid, 25, 29
Cathode blanks, 62, 103, 116, 216, 223
Cathodes:
  ceramic, 179, 191 (*see also* Anodes,
    ceramic)
  chromium containing film, 217–218
    Auger-spectrometric investigations,
    215
  nickel-molybdenum coating on stainless
    steel, 191
  oriented, 177–178
  platinized titanium, 177–178
  palladium, 178
  platinum, 178
  with stagnant zones, 210

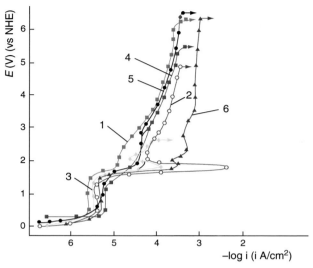

Plate 1 Anodic polarization plots in 1 g/l NaCl on titanium (1) and on its alloys with: 2 – 3% Al; 3 – 6% Al; 4 – 5% Ta; 5 – 0.2% Pd; 6 – 2% Ni. Temperature, 90°C (See Figure 6.15, p. 98).

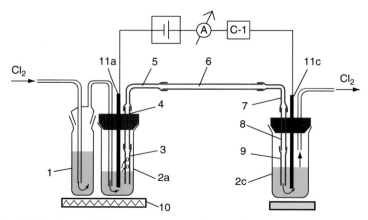

Plate 2 Installation for modeling operation conditions of titanium piping for wet chlorine in the field of external current (explanations are given in the text) (See Figure 9.2, p. 147).

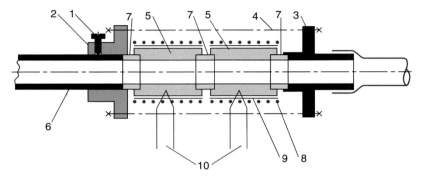

Plate 3 Device for condensate layer interruption (explanations are given in the text). (See Figure 9.5, p. 149).

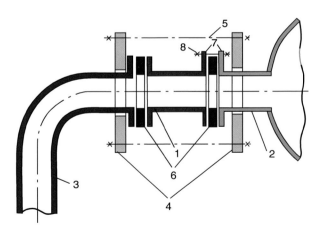

Plate 4 Protection of a branch of titanium group header for wet chlorine by titanium sacrificial anode for leakage current drainage (explanations are given in the text) (See Figure 9.8, p. 156).

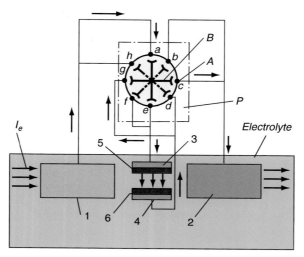

Plate 5 Simplified scheme of a device with two alternative electrodes for protection of metallic structures from corrosion by external currents under the conditions of electrolysis with metal deposition on the cathode (explanations are given in the text) (See Figure 9.11, p. 159).

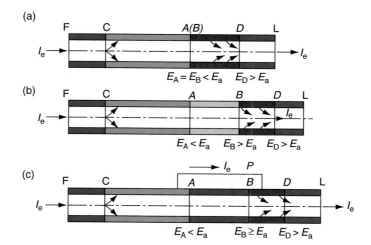

Plate 6 Protection from corrosion by the attack of an external current on tubes filled with electrolyte by using anodes: a – extended; b, c – remote, placed towards the direction of the external current field. BD – anode; AB – interstitial tube of conductive (b) or insulating (c) materials; FC and DL – parts of insulating tubes; P – external conductor (See Figure 10.4, p. 171).

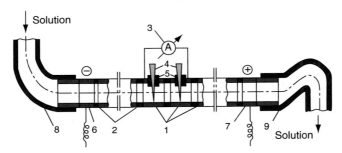

Plate 7 Model of a compound tube for estimation of corrosion protection efficiency by anodes oriented in the field of the external current (explanations are given in the text) (See Figure 10.5, p. 172).

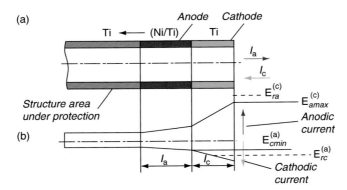

Plate 8 Potential distribution (b) along the composite electrode (a), placed along the field of external currents of anodic (I_a) or cathodic (I_c) directions (See Figure 10.9, p. 180).

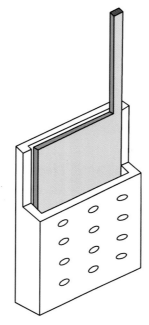

Plate 9 Perforated Teflon case specimen holder (explanations are given in the text). (See Figure 11.15, p. 211).

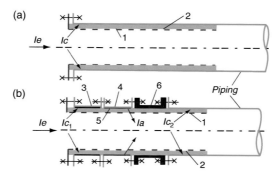

Plate 10 Design of a stable titanium cathode with a stagnant zone for protection of titanium piping from corrosion by cathodic external current in media containing: (a) – dissolved chlorine and chlor ions; (b) – chlor ions only (explanations are given in the text) (See Figure 11.17, p. 212).

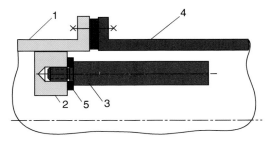

Plate 11 Mounting of rod-shaped copper anode inside a protected pipe (explanations are given in the text) (See Figure 12.1, p. 222).

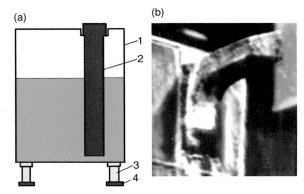

(a)

(b)

Plate 12 Titanium degassing tank (1) with protection from corrosion by anodic leakage current with the help of dissolving anode (2): **a** – design of the tank with anode inside (explanations are given in the text); **b** – photograph of the upper part of the tank with a slit in which an ear of the anode is introduced (See Figure 12.3, p. 224).

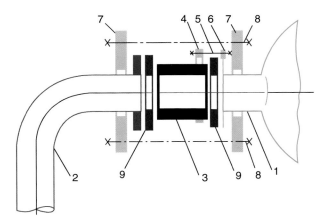

Plate 13 Protection of titanium branch tubes of group headers for wet chlorine with the help of current leak-off anodes of graphite (explanations are given in the text) (See Figure 12.4, p. 225).

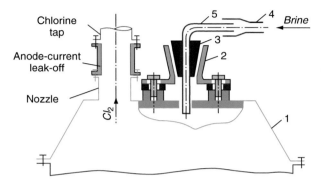

Plate 14 Protection of nozzles for connecting chlorine taps to headers and for brine input mounted on titanium covers of diaphragm electrolyzers (explanations are given in the text) (See Figure 12.5, p. 226).

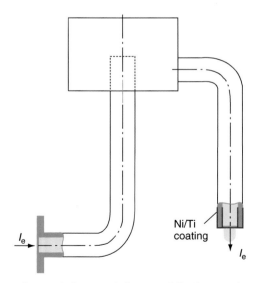

Plate 15 Titanium overflow unit for caustic liquor of diaphragm electrolyzer provided with protection from electrocorrosion by Ni/Ti coating obtained in an electric spark explanations are given in the text (See Figure 12.6, p. 227).

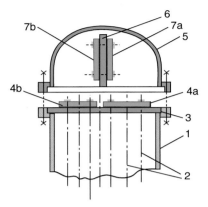

Plate 16 Protection of titanium shell-and-tube heat exchangers (brine heaters) by current leak-off anodes (explanations are given in the text) (See Figure 12.8, p. 231).

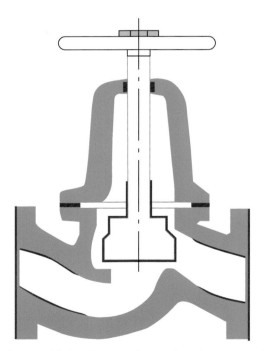

Plate 17 Titanium valve provided with protection against electrocorrosion by ORTA coating indicated by red lines (See Figure 12.9, p. 232).

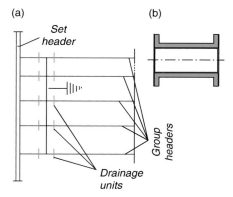

Plate 18 a – Scheme of installation of units for current drainage from brine to earth for group brine headers; b – unit of titanium with ORTA coating indicated by red lines (See Figure 12.10, p. 233).

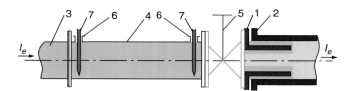

Plate 19 Scheme of protection of stainless steel 18-10 piping in technological solution of copper electrorefining with the help of a lead anode oriented in the field of the external current (explanations are given in the text) (See Figure 12.11, p. 233).

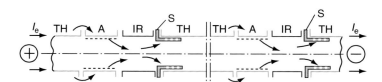

Plate 20 Combined protection of titanium chloranolyte header from corrosion by external currents of anodic and cathodic directions: TH – protected header; A – current leak-off anode with ORTA coating (indicated by dotted lines); IR – insulating inserts; S – perforated sleeves of Teflon. Arrows indicate the current direction (See Figure 12.12, p. 236).